SYMBOLIC COMPUTATION

Managing Editors: J. Encarnação P. Hayes

Artificial Intelligence
Editors: L. Bolc A. Bundy J. Siekmann

Springer Series
in Symbolic Computation

Editors

Computer Graphics: J. Encarnação; K. Bø, J.D. Foley, R. Guedj,
 J.W. ten Hagen, F.R.A. Hopgood, M. Hosaka, M. Lucas, A.G. Requicha

Artificial Intelligence: P. Hayes; L. Bolc, A. Bundy, J. Siekmann

---

M. M. Botvinnik

Computers in Chess
Solving Inexact
Search Problems

Translated by Arthur A. Brown

With Contributions by A. I. Reznitsky, B. M. Stilman,
M. A. Tsfasman, and A. D. Yudin

With 48 Illustrations

Springer-Verlag
New York Berlin Heidelberg Tokyo

M. M. Botvinnik
c/o VAAP—Copyright
Agency of the U.S.S.R.
B. Bronnaya 6a
Moscow 103104
U.S.S.R.

Arthur A. Brown (*Translator*)
10709 Weymouth Street
Garrett Park, MD 20896
U.S.A.

Library of Congress Cataloging in Publication Data
Botvinnik, M. M. (Mikhail Moiseevich), 1911–
   Computers in chess.
   (Symbolic computation. Artificial intelligence)
   Translation of: O reshenii netochnykh perebornykh
zadach.
   Bibliography: p.
   Includes index.
   1. Chess—Data processing. 2. Search theory.
I. Title. II. Series.
GV1447.B67513   1983   001.4'24   83-10571

Original Russian edition: *O Reshenii netochnukh perebornykh zadach*. Moscow:
Nauka, 1978.

Typeset by Science Typographers, Medford, NY.
Printed and bound by R. R. Donnelley & Sons, Harrisonburg, VA.
Printed in the United States of America.

9 8 7 6 5 4 3 2 1

ISBN 0-387-90869-2 Springer-Verlag New York Berlin Heidelberg Tokyo
ISBN 3-540-90869-2 Springer-Verlag Berlin Heidelberg New York Tokyo

Preface to the English Edition

Much water has flowed over the dam since this book went to press in Moscow. One might expect that PIONEER would have made substantial advances—unfortunately it has not. There are reasons: the difficulty of the problem, the disenchantment of the mathematicians (because of the delays and drawing out of the work), and principally the insufficiency and sometimes complete lack of machine time.

The general method used by PIONEER to solve complex multidimensional search problems had already been formulated at that time. It was supposed that the successful completion of the chess program PIONEER-1 would provide a sufficient validation for the method. We did not succeed in completing it. But, unexpectedly, PIONEER's method obtained a different kind of validation.

Since our group of mathematicians works at the Institute for Electroenergy, we were invited to solve some energy-related problems and were assigned the task of constructing a program that would plan the reconditioning of the equipment in power stations—initially for one month. Until then, the technicians had been preparing such plans without the aid of computers.

Although the chess program was not complete even after ten years, the program PIONEER-2 for computing the monthly repair schedule for the Interconnected Power System of Russian Central was completed in a few months. In mid-October of 1980 a medium-speed computer constructed the plan in 40 seconds. When, at the end of the month, the mathematician A. Reznitsky turned over the results to the Central Dispatch Control (CDC) of the power system, he was treated with disbelief, since the plan already prepared by the technicians differed from the computed plan. In a day or so, however, things were cleared up. PIONEER-2 turned out to be more competent than the humans. Using the methods of the chess master, the computer very quickly found a high priority variation in the plan, tested the possibility of improving it, and produced the results. PIONEER-2 was at once adopted by the CDC for implementation.

In the following year, PIONEER-3 was developed to produce the annual plan for all power stations in the USSR. The plan for 1982 was produced in 3 minutes 19 seconds. If one notes that the monthly plan dealt with 200 units for 30 days, and the annual plan with 600 units for 365 days, one must be amazed; the dimension of the full-width search tree for the annual plan is essentially infinite. The truth of the matter is that by using the chess master's method, the search problem is reduced to one of analysis, and therefore the solution depends only weakly on the dimensions of the search.

In 1982 the program was updated. It not only produces the plan, but if necessary minimizes the increase in the reserve power that must be dedicated to offset the output of the units in repair. The technicians like this very much, since now they can only approximate the amount of reserve power needed for maintenance; the computer itself made the value of the reserve more precise. However, the program was more complex and the 1983 plan consumed 12 minutes 6 seconds.

Why should the maintenance planning present a simpler problem than chess? The answer is not hard to find. Let us look at two schemes for solving an enumerative problem.

Scheme (a) corresponds to a solution of the problem by a full-width search. It is a simple scheme, but suitable only for the case in which the branching factor during the search is small; only then can we obtain a deep solution. For a branching factor appreciably greater than unity we can in general obtain only a weak and superficial solution because of the catastrophic growth of the search tree. Moreover, and this is the essential point, since the full-width search is not connected with the essence of the problem we are trying to solve, a good positional estimate is excluded; without it we cannot find a good solution.

The chess master uses Scheme (b). He processes his initial information, establishes a goal for the inexact game, establishes a multi-level system, sets priorities for the inclusion of moves for consideration, and constructs a positional estimate. After this, the game of chess—a search task of very high dimension—reduces to a problem in analysis; the branching factor remains close to unity, and nothing prevents reaching a deep solution.

We can now see why maintenance planning is easier than chess. In the planning problem, the initial information fed to the computer scarcely needs processing; it is already in a form suitable for analysis. In chess, on the other hand, the data destined for analysis is deeply hidden in the initial data. The principal task consists in transforming the initial data to a form suitable for analysis. Herein lies one of the reasons for our delay in finishing PIONEER-1.

Nevertheless, the chess program has made some progress. Where before we looked on chess as a three-level system (attack trajectories with attacking and attacked pieces, fields of play, the ensemble of fields) we now model the game of chess as a four-level system. A field of play has a somewhat abstract nature; on the basis of the field we have now formed a real chain of trajectories (this is the third level) and an ensemble of such chains (the fourth level) which is a genuine mathematical model of a position.

We had already developed the concept of the compound field, composed of a number of simple fields, but we did not know how to analyze it. The priority for inclusion of moves in the search was based on the "practicability" of the several trajectories, and such a priority did not yield good results. We now base the priority on the practicability of a chain of trajectories, which we call a compound field. To a first approximation we may say that the trajectories in a chain belong to two fields. A chain must have its own basic attack trajectory and, of course, the target of attack. As we noted above, an ensemble of chains constitutes the mathematical model.

The positional estimate is now based not only on material values but also on the situational value of the pieces. The concept of the situational value had already been introduced in the author's earlier book *Computers, Chess, and Long-range Planning*, but it was not formalized. We have now succeeded in doing that. The greater the value of a chain (of trajectories) with which a piece is connected, the higher the situational value of that piece.

This was tested on a position in a game by Botvinnik–Capablanca. We succeeded for the first time in increasing the positional estimate in the course of a sacrificial combination. We are currently sharpening some new developments, after which PIONEER will be suggested for the analysis of quiescent positions.

Few people believe in the success of our work. Nevertheless, I had not expected Ken Thompson to be skeptical; so far as I know, Claude Shannon is also skeptical. This is most curious, since in the historical development of an artificial chess master there have been only two major events: the fundamental work by Shannon (1949), and the construction of BELLE, a high-speed specialized computer by Thompson (1980). BELLE has attained national master rating and is World Champion among chess-playing computers. However, BELLE uses the brute force method, and this is hardly capable of further progress. It is the computer's turn to adopt a more fruitful method—perhaps PIONEER. And if PIONEER is unsuccessful, we must believe that some other method will be found. The problem must and will be solved.

Note: Recently the solution to the maintenance planning problem has again been advanced. The program PIONEER-5 will be completed in December. It will deal with a whole set of resources expended in the maintenance process, instead of with one resource only. Since these resources are in part local and in part centralized, PIONEER will begin with local preliminary plans, for orientation, and then proceed to the second and higher levels. It will then reverse the process and return finally to the lower levels, where priority will be given to the general interests of the integrated energy system; the local plans will then be optimal.

After PIONEER-5 has successfully completed its trials, one may assume that, to a first approximation, it will be capable of planning any branch of the economy.

As for PIONEER-1, there remains the completion of the positional estimate, and then further progress can be made.

Moscow   M. M. BOTVINNIK
June, 1982

Preface to the Russian Edition

This book gives an account of the theory needed for the solution of inexact enumeration problems; the theory as expounded here is to some extent based on hypothesis, since our experience does not yet fully support our theoretical position. When our chess program PIONEER begins to play at master strength, we may say that the theory has a solid basis.

The (unfinished) history of the development of strong chess programs is connected with a struggle between two different trends. The prevailing opinion, for a long time, was that the computer should not imitate a chess master's thought processes, and that the method for play by a machine should be based on an exhaustive search for possible moves. Since the first successes of PIONEER, the position has changed to some extent; from now on, computer programs will increasingly tend to imitate humans.

The first part of the book contains a general statement of the method that, in our opinion, should be used for the solution of inexact enumerative control problems; we use the game of chess as an example to show how the general theory can be successfully applied. A detailed exposition of the algorithmic basis is given in the appendices, which were written by mathematicians who took part in the development of PIONEER. They should be of interest to program designers and should aid in the practical application of the principles set forth in this book.

Contents

APPENDIX 4

An Associative Library of Fragments (by A. I. REZNITSKY AND
    A. D. YUDIN)

CHAPTER 1
The General Statement

Definition of an Inexact Task

The notion of an inexact problem was introduced by the author several years ago [1], but no precise definition was then given. We now say that an enumerative task is inexact if it solves a problem by minimax methods on a truncated search tree (see the Glossary of Terms). The concepts of minimax procedure and search tree are well known; the concept of the truncated search tree may need explanation. Problems soluble by the formation of a tree of all possibilities (elementary actions) may differ in difficulty and may give rise to search trees of different sizes—small, large, or even infinitely large. If the resources of our information-processing device (speed and memory) are so limited that we cannot form the tree and search it exhaustively, we must either abandon the task or be content with an inexact (i.e., approximate) solution. If an inexact solution is acceptable, we limit the depth of the variations; the use of a depth-truncated tree and the acceptance of an approximate solution make the enumerative problem inexact. The definition of an inexact problem is therefore inextricably bound up with the general method of solution and with the resources of the information-processing system being used. If we can apply the minimax procedure to the complete tree, the problem is exact. (Problems may of course be solved exactly by other methods, e.g, by the use of equations or exact algorithms, and they may be inexact under other definitions, as when approximate solutions to equations are used. However, we shall use the narrow definition of an inexact task, as given above, and will not consider inexact tasks not conforming to it.

Inexact Tasks and Control Systems

A control system has three functions: (1) the receipt of data, (2) the processing of data, and (3) the execution of the resultant solution. The data-processing component may have tasks of varying complexity and difficulty, depending on the assignment given the control system. When the system solves its task by forming a search tree, the theory of inexact problems and the theory of control are inseparable.

Although at first glance the theory of inexact problems seems abstract, it has, in fact, great practical significance, since most control tasks are inexact. We investigate the theory by applying it to the game of chess, which represents a typical inexact task.

Two Methods for Solving Inexact Problems

Two essentially different methods for solving inexact problems were described by Shannon [2] in 1949, when he posed the problem of designing a chess-playing program. In the first method, all possibilities (moves) are included in the truncated search tree; there are no exclusions. In the second, moves that are known to be senseless (cf. Glossary of Terms) are excluded, so that the tree consists of potentially sensible moves only. The first method and its implementation are in principle simple; Shannon recommended its use, and it has been applied by mathematicians. He noted that the second method offers better prospects, but outlined no approach to its implementation.

The first method, however, is hopeless as a means of finding a good solution to an inexact task, as we shall show by applying it to the game of chess. On the average, each side in a chess game has 20 available moves. Suppose we start in a position where White is to move; all 20 moves of White's pieces must be included in the search tree. Then it is Black's turn; for each of White's 20 moves, Black has 20 answers, so that the tree contains 400 Black moves. If we extend the variation to include another move by White, the tree will contain a total of 8420 nodes. If the variation is to contain three complete moves—six half-moves ["plies"]—it will contain some 67 000 000 nodes!

The size of the tree is an exponential function of the depth. Thus the tree grows catastrophically (see Fig. 1, where, as is customary, the tree is shown growing downward from its root). We are to find the optimal variation in this tree; it may easily consist of six moves. Then we have more than enough moves in the tree; we are looking for a single needle in a haystack. Worse yet, in the search for a needle we at least know what we are looking for,

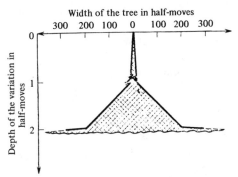

Figure 1   Growth of the search tree in a chess game with an exhaustive search.

whereas in the search for the optimal variation this is often far from being the case.

If we develop a suitable algorithm for a chess program, the resources of the computer (memory and speed) are less significant than might be supposed at first glance. Whenever we extend the depth of the variation by a move, the resources of the machine must increase by several factors of 10; this is impossible in practice. Thus the main element is not the power of the computer, but the skill of the programmer, i.e., the extent to which he has contrived to depart from the fundamental principle (a full-width search) in order to economize on machine resources. This fact explains how the Soviet program KAISSA won in the First World Computer Chess Championship at Stockholm in 1974, even though the computer used has comparatively modest capabilities. There were competitors with several times as much power, but they were defeated by KAISSA. An increase in computer power is of little help in solving inexact enumerative problems. In a quarter century the computer has been raised to second class rating as a chess player; in the last five years no progress has been made.

Two tasks confront the programmer in his attempts to perfect his program: improve the optimal variation (find a stronger one), and accelerate the solution of the search problem. The acceleration is necessary if the optimal variation so far found is not good enough, since then the depth of variations must be increased, with a consequent increase in the volume of the tree. Again using chess as an example, let us note that this line offers very limited possibilities for success if the program is based on an exhaustive enumeration of the possibilities (full-width search).

Suppose that on the average, in any position, m moves are available. Then for a given depth consisting of n plies and for a full-width search, the number of nodes in the tree will be $A_n = m + m^2 + m^3 + \cdots + m^n$. By the well-known branch-and-bound method (in the West this is also known as the α-β-cutoff method), this can be reduced to the value $A'_n = A_n^{1/2}$. If we set $m = 20$ and $n = 6$ (these are reasonable values), we have

$$A'_n = \sqrt{20 + 400 + 8000 + 160\,000 + 3\,200\,000 + 64\,000\,000} \approx 8000.$$

In practice, the branch-and-bound method leads to no such reduction in the number of nodes. For instance, the program CHESS 4.6, with a limiting depth of variations set at 6 plies, has a search tree containing some 400 000 nodes; this differs significantly from the estimate A'_n. (In fact, the search tree contains some forced variations, relating to captures and mates, but their influence on the number of nodes is small.) Let us therefore take $A'_n = A_n^{2/3}$ rather than $A_n^{1/2}$. Then we find 160 000 nodes in the search tree instead of 8000; this is considerably closer to the truth.

We now display the number of nodes in the tree as a function of the limiting depth of a variation, assuming that the branch-and-bound algorithm is not used. We take two cases, $m = 20$ and $m = 7$. Inspection of the curves in Fig. 2 will show that the limiting depth of a variation cannot be increased beyond $n = 5$ or $n = 8$, and that the dependence on the average number of moves available is weak. Since the number of moves that can be searched may be taken as proportional to the speed of the machine, it is clear that an increase in speed, beginning at some given depth of variations, cannot strengthen the play. If the number of moves available at each node is decreased to $m = 7$, the situation is somewhat better: an increase in speed, yielding an increase in A'_n, has a greater influence on the limiting depth n of the variations. The authors of CHESS 4.6 took advantage of this fact in programming the endgame, when the number of pieces is decreased and therefore m is decreased. Using a Cyber-176 computer, with a speed of twelve million operations per second, they succeeded in deepening the variations from a length of 6 to a length of 12 and obtained stronger optimal variations.

If, on the other hand, a strong game has already been obtained, and the object is merely to increase the speed with which the solution is reached, then increasing the speed of the computer does solve the problem completely. This completes our summary remarks concerning the first method, the full-width search.

The prospects for the second method are founded first of all on the fact that it is the one used by humans, in particular by strong chess players. By excluding obviously senseless moves from the tree being formed, a human seeking an optimal variation searches a deep and narrow tree. In this case, the number of moves in an optimal variation is comparable to the number of moves in the tree.

Some mathematicians, while admitting that this is the pattern of human thought, nevertheless contend that a computer should act differently. They say human thought is not an optimal model for computers and there is no point in transferring such modes of thought to a computer program. If there is merit in this argument, we may ask: Why then over a quarter of a century have the efforts to teach a computer to play chess by non-human methods yielded such weak results? If in fact chess programs based on a full-width search and truncation of the search tree were to beat chess masters, we might well say that a computer should act along lines differing from human thought processes—but up to now computer programs have not beaten chess masters.

In support of the view that computers should behave like computers, some specialists put forward the example of heavier-than-air machines—they note that airplanes and birds fly in different ways. From the mechanical point of view, it is true: airplanes use engines for motive power and birds use wings. But from a deeper point of view, their tasks are the same: both airplanes and birds overcome gravity and air resistance. One might say by analogy that both men and computers, in solving an inexact task, must overcome the combinatorial explosion of the task by forming a deep and narrow tree. Men have long known how to do this; it is time to transfer to computers the results of thousands of years of experience. If this is done, and chess programs succeed in forming deep and narrow trees, then probably the power of the machine will play a role, and the most powerful machine will defeat its opponents.

In conclusions we may ask: Which method, and under what circumstances, should we prefer? Speaking generally, we may make the following observations:

Inspection of Fig. 2 shows that when the number of moves at a given node (in a given position) is decreased, the curve $y = f(x)$ shifts to the right, where y denotes the number of nodes in the truncated tree and x denotes the limiting depth of a variation; then the full-width search method yields a deeper solution. When the number of moves at a node increases, the curve shifts to the left; the depth of a variation decreases, and the solution becomes weaker.

Thus, in complex control problems, where the number of possibilities under consideration is large, the full-width search is of doubtful value, but it can be recommended in simple cases, since the second method is more complex. The second (human) method is especially valuable in complex

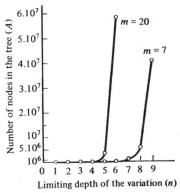

Figure 2 Dependence of the number of nodes in the search tree on the limiting depth of a variation.

cases, but can hardly stand up to competition from the full-width search method in simple cases.

The Goal of the Game and the Scoring Function

The goal is the basis for the fundamental algorithm of a game (for the solution of an inexact problem); for instance, the goal of the exact game of chess is to checkmate. However, in a truncated tree the exact goal plays almost no role, since no variation leads to it. We must introduce a new (intermediate) goal, an inexact goal for the inexact task corresponding to the truncated tree (Fig. 3). (We recall that with a truncated search tree, an arbitrary problem—even if finite—becomes inexact.)

How is the goal used to find the solution of a task? How does it influence the formation of a search tree? An inexact goal yields a logical basis for breaking off variations in the tree; a variation is pursued to its limiting depth only when pursuit of the goal dictates that it should be carried so far or further.

If the goal is reached, or if we find that it cannot be reached, the variation is broken off; this may happen before the limiting depth set by the truncation is reached.

In all existing chess programs, however, the variations are pursued to some limiting depth established so as to correspond to the truncation of the

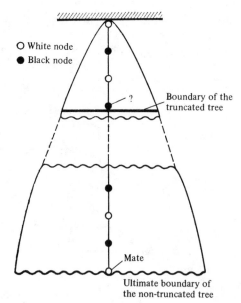

Figure 3   The exact and inexact goals of the game in chess.

tree (and if some are pursued further they are so only up to some other pre-established limit). The pursuit of all variations to their limiting depth is a sure sign (without exception!) that in the corresponding algorithm there is no inexact goal for the inexact game. This is a striking fact that should be especially noted: in these programs the goal of the game (mate) does exist in implicit form (only when mate is reached is a variation broken off short of the limiting depth), but mate, as an explicit goal, belongs to the exact game only.

Shannon was inconsistent when in 1949 he proposed to teach a computer to play chess with the aid of a truncated tree. This was a great step forward, but—because of his authority—in establishing the goal as that of the exact game and not completing the second necessary step, he doomed a number of mathematicians to a truly "aimless" labor. There was, of course, some usefulness in these mathematical researches, but only this—they succeeded in convincing the mathematicians that without an inexact goal it is impossible to solve an inexact problem.

Next, what is the role of the scoring function? If there is no inexact goal, there is only one use for the scoring function: it evaluates a variation when it is broken off. If there is an inexact goal, the scoring function is used in deciding when to break off a variation and, as we shall see later, it is used to develop a deep and narrow tree. It preserves its ordinary basic purpose: to assess the variation obtained by a minimax procedure. We repeat: in the absence of an intermediate goal, the scoring function serves only to evaluate a stopped variation.

The goal of the game provides an answer to the question. What are we aiming for? The scoring function marks the extent of our success in reaching the goal. Shannon's algorithm contains a scoring function, but no goal. Therefore, in Shannon's method we can assess a variation and compare it with other variations; but since we have no goal we cannot rationally break it off short of its limiting depth.

The goal and the scoring function represent different concepts. For a fully valuable solution of an inexact task, both should be formalized and used.

Goal and Prognosis (The Optimal Variation)

We obtain an optimal variation as a result of a minimax procedure over a truncated search tree. We should consider an optimal variation and its value as a prognosis of the extent to which an inexact goal can be realized. Neither the optimal variation nor its assessed value are (nor can be) the goal of a game. The optimal variation is only a momentary forecast of the attainment of the goal. This forecast is provisional, in that fresh information may bring changes. Everything depends on the situation.

Changes in the optimal variation are primarily connected with the truncation of the tree. Some of the variations may have reached the limiting depth, and so were broken off not for logical reasons but under compulsion. The values of these variations are known, but are they reliable? In the course of time, we move along the optimal variation and must continue to reassess it if the limiting depth remains unchanged. It may cease to be optimal. Here is one of the sources of inexactness in the task—the inexactness of the forecasts. The type of the goal remains fixed, but its value and the forecast are variable.

The value of the optimal variation may be regarded as a forecast for inverse-feedback purposes. Let us assume that a control system is acting along an optimal variation, to the extent that this variation has been computed. Then the assessed value of the variation as realized is a real inverse-feedback signal. When we are in the initial position, however, i.e., when the optimal variation has just been established and the control system has not begun to act, the value of the variation is a forecasted inverse-feedback signal.

We repeat: the value of the optimal variation (the forecast) can in no way be equivalent to a goal of the game. The value may be either good or bad, and an undesirable value cannot be a goal for a control system.

Multi-level Control Systems

A control system may consist of a set of subordinate control systems; this set may form part of a new set. Fig. 4 displays a three-level control system.

Each level must have its own individual goal, but the goals at the several levels must be all of the same type; otherwise a multi-level system cannot function successfully. If at some level the goal differs in kind from the goals of other levels, the given level will form an alien element in the multi-level system.

In this context, the notion of "level" can be formalized, and must be. We define a level in a control system as a subsystem that has its own goal (though typical of all goals in the system) and in which the search is stopped when the goal is attained or is known to be unattainable.

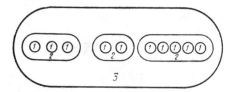

Figure 4   Three-level control system. 1, 2, and 3 are the first, second, and third control levels, respectively.

This definition has a decisive influence on the formation of levels, and points up the highly important role of the goal in the formation of a multi-level system.

For the formation of deep and narrow trees, multi-level systems offer great advantages as compared to one-level control systems. The essential advantage is that when we are forming a search tree and the situation at some level has been clarified (the local goal has been reached or has been found unreachable), the local search is stopped, though it continues at other levels. Later we shall note other merits.

Multi-level systems are widely encountered. A factory, for example, may be considered as a four-level control system: the worker, the section, the department, the factory. If a section has done all of its assigned work in half a working day, and there is nothing left to do (the goal is attained or is unattainable), the department head responsible for production will look for an optimal variation, excluding the section level from his search tree.

Types of Multi-level Systems

Multi-level systems may differ in the distribution of information-processing resources among the several levels, that is, in the location of the decision-making elements. The various types are as follows (the labelling is non-alphabetic in the order of its presentation, but ultimately makes sense):

(A) Control is vested in a single center, with a single information processor. The game of chess provides a typical example: although we shall see later that chess may be regarded as a three-level system, we note that the control of all three levels still lies in a single center—the mind of the player.

(B) Each level has its own information processor and its own control center. The factory example described above will illustrate this type.

(F) Only the lower levels have control centers, and the ensemble of these centers is uncontrolled. This type is the opposite of type (A). It is characteristic of inanimate nature and the vegetable kingdom. We may say that in this type the lower levels do not form an assembly, but rather a collection. At the constituent levels, only egoistic interests exist; there is no overall goal at the level of the collection.

The difference between types (A) and (B) is not only one of principle; there is also a difference in the required apparatus. The presence of a set of control centers [type (B)] is typical of a computational complex or an electronic multi-processor. Thus, when a task can be executed in parallel, the existence of a set of control centers leads to an increase in speed and memory of the information processor. The increase can be very great, and therein lies the advantage of the type (B) multi-level systems.

If several information processors (human or electronic) solve the same problem, the task will be accomplished only at the speed and depth belonging to the most powerful among them. Of course, the remaining processors do not necessarily waste their labor, since they may contribute to the elimination of errors, but that is all! If the same processors solve different component problems making up a whole, then the total problem will be solved as though by a processor having the sum of the capabilities of the individual components.

Thus the advantage of the multi-level control system lies not only in the fact that the search may be cut off at various levels, as soon as the question of reaching their goals has been settled, but also in the fact that a multi-level system can serve as a set of processors and thus increase the speed and depth of the solution.

When a multi-level system has a single control center, as in chess, only the power of a single information processor is used. It would be difficult to convert chess into a system with a set of centers, since, as we shall see later, the levels in chess are not fixed but are in continual change. In other multi-level systems, however, as in the factory, the control system may use the power of many processors, since the levels in such a system are fixed and separate processors can be used for control.

As we have already remarked, the several levels must have homogeneous, though individual, goals. The use of processors at different levels requires essentially that the control centers be autonomous. It is necessary to prescribe the service to be performed by each of the goals, to program it, and to make it profitable for the corresponding level; the rest can be left to local autonomy. Only if this is done can the use of a set of information processors be advantageous), and only then can we obtain the corresponding increase in speed and memory. If a solution achieved at one level is not used by the immediately superior level, resources will have been expended wastefully.

We have considered three types of multi-level systems [(A), (B), and (F)] which are distinguished by the distribution of the control centers among the several levels. We now classify the type (B) systems with respect to the priorities (for decision making) of the goals in the various levels.

Each level has its local egoistic goal; but the local goal of the supreme level, that of the ensemble, appears as a common goal for its subordinates. Only the local goal of the highest level appears as the common goal of the whole multi-level system, the goal for the control systems at all levels. If a decision is made in accordance with local goals, the result will be an optimal variation from the viewpoint of the local interests.

Making a decision based on local interests does not mean that the overall goal is necessarily ignored completely, since the total interest may coincide with the local interests. If the decision is made on the basis of the overall interest, the local interests are not necessarily denied, since they may coincide with the common goal. The main question is: Do local interests or common interests prevail in the decision-making process?

In the programs at each level there is a duality of goals—those of the local level versus those of the next higher level. There are two types of priorities, and we may further distinguish systems of type (B) in accordance with them:

(C) Systems in which the control programs at the various levels give priority to local interests. The interests of the higher levels are taken into account only when they are not in conflict with local interests.

(E) Systems in which the control programs give priority to the higher goals. Local goals are taken into account only when they do not conflict with higher-level interests.

Advantages of the General Goal

If in a multi-level system the common goal has priority [type (E)], the value of the optimal variation will be higher than if the programs of the various levels give priority to local goals [type (C)]. In chess, this assertion can be supported by practical experience. In general, the position stated is well known. Krylov's fable of the Swan, the Crayfish, and the Pike will illustrate it: Each of the three proposed to pull a boat, but the Swan chose to pull it into the air, the Crayfish onto the beach, and the Pike under the water! (An academic cyberneticist recently stated that chess is much more complex than politics or economics. This is not so. Chess represents a multilevel system with a single control center and so is less complex than a system with many control centers; the systems characteristic of politics and economics are of the latter kind.)

Suppose for simplicity that our system has two levels and is of type C. Let us denote the components of the first level (of which there are k in number) by the indices $1,\ldots,i,\ldots,k$, and denote the total system by the index S. We shall show that if we go from a control system of type C to a system of type E, we may guarantee not only to reach at least the value of the optimal variation under the C regime, but also to increase it.

We denote the values of the optimal variation or S under the C regime by D', and under the E regime by D''. Our assertion states that $D'' > D'$; we write the difference as $\Delta D = D'' - D'$. Under the C regime, the values of the optimal variations of the individual first-level components will be, say, $d'_1,\ldots,d'_n,\ldots,d'_m,\ldots,d'_k$ and the sum of these is $d'_1 + \cdots + d'_n \cdots + d'_m \cdots + d'_k = D'$.

Let us now control the system S by method E. Then we obtain the optimal variation for the whole system with value D'' equal to the sum of the values $d_1'', \ldots, d_m'', \ldots, d_n'', \ldots, d_k''$ of the individual optimal variations in the lower-level components. Let us assume that the values $d_1'', \ldots, d_m''$ are larger than the corresponding values $d_1', \ldots, d_n'$, that the values $d_{n+1}'', \ldots, d_m''$ are equal to the values $d_{n+1}', \ldots, d_m'$, and that the values $d_{m+1}'', \ldots, d_k''$ are less than the corresponding $d_{m+1}', \ldots, d_k'$. Then we reduce the values $d_1'', \ldots, d_n''$ to the corresponding values $d_1', \ldots, d_n'$, leave the values $d_{n+1}'', \ldots, d_m''$ unchanged, and increase the values $d_{m+1}'', \ldots, d_k''$ to the values $d_{m+1}', \ldots, d_k'$. Thus we obtain the forecast for the control system ignoring the local interests as

$$D'' = D' + \Delta D = d_1' + \cdots + d_k' + \Delta D.$$

Let us consider the case in which the optimal variation has been precisely defined. We have the value D'' for the regime E. Expending the amount $D' = d_1' + \cdots + d_k'$ for distribution to the components $1, \ldots, k$, we still have the amount ΔD as a supplementary amount for distribution.

Thus, it is possible for a system of type E to yield a higher value for the variations of the components. Nothing needs to be changed in the system except the control system, as we shall immediately show.

The Method for Connecting the Optimal Variations of the Components for Types C and E Regimes

We continue to consider a two-level control system S. Let us single out one of the lower-level components i. Under the regime C, it will assign a value d_i' to its optimal variation, and under E, it will assign the value d_i''. We shall suppose that $d_1' \neq d_i''$, and first consider the case in which $d_i' > d_i''$. Naturally, because of the predominance of local egoistic interests, the regime C disregards the control system E and the values d_i'' connected with it, since the system i would gain less value under E. How do we change the control programs of the system i and the system S so that we obtain the desired result?

Since nothing prevents the system i from seeking maximum profit, we must impose a supplementary limit d_i' on the profit it can obtain. In order to prevent it from securing the surplus benefit $d_i' - d_i''$, we must revalue the costs of the material resources for this component only (and in the present case, increase them). In chess this amounts to increasing the value attached to the pieces taking part in the action under the local control system (for example, in a field). The increase in this price list should be such that in the new calculation the value d_i'' will be equal to the value d_i' in the old calculation. With these two changes (limitation of the profit to the value d_i' and the change in the price list), the goal of the local egoistic program is no longer in conflict with the overall goal of the system S. The first change, limitation of the profit, refers to the system i; the second, the change in the price list for i, refers to the system S.

When $d_i' < d_i''$, the changes in the price list should be in the opposite direction; if $d_i'' < 0$, they must involve a change of sign. With these changes in the local and overall programs the system S will be able to realize the profit ΔD under the regime E without breaking its principle, i.e., without changing the character of the goals of the egoistic programs at the component level.

Computer Programs and Humans

Up to now we have been supposing that the control centers use computers. But today we know that the human mind is the foundation of the control center. What are the characteristic differences between humans and computers in control systems?

From the cybernetic viewpoint, the difference is easily formalized. A computer program, generated by a programmer, is arbitrary (within the limits set by the resources of the computer). The human cannot be programmed in such an elementary fashion; his program is formed slowly, by interaction with the outside world, using a process of self-teaching.

Changing the program of a human is a complex act. All the same, it is possible to influence it, by changing the reaction of the outside world, by using inverse feedback, and by setting up a program of self-teaching.

There are many examples: For instance, a lazy recruit is taken into the army; he remains lazy, but his laziness does not manifest itself, since he is fully aware of the consequences if he neglects his military duties...A driver may not object to drinking, but while on the transport base he is a teetotaler, since the punishment for drinking would be heavy...A young mother abstains from her passion for theater-going, since her baby is ill and she has no one with whom it can be left...An engineer believes that his superior has made a wrong decision, but he avoids a dispute, fearing that he will lose an attractive job...etc.

A computer program cannot be egoistic, from the viewpoint of the computer, or at least only to the extent that it takes care of the local interests of the system that it controls. The program of the human may include self-interests in conflict with the interests of the total system he controls, and this fact may have a negative influence on the control system.

In a computer the selection of a decision and its adoption amount to the same thing. These two actions may be different in the human, who may select a decision corresponding to the interests of the total system, yet adopt one corresponding to his egoistic interests. Thus, from the cybernetic viewpoint, an artificial intelligence is preferable to a natural one. The problem lies only in the fact that up to now we have no sufficiently powerful artificial intelligence.

The Expansion of Artificial Intelligence

If the executive organ of a control system is too powerful, we may not want to expand the system (we are thinking of nuclear weapons and nuclear energy in general) since the greater the number of countries that possess it, the higher the probability of misuse. With respect to information processing organs, however (e.g., artificial intelligence), the problem is to accelerate their expansion. We must keep in mind that the problem of non-proliferation of nuclear weapons lies solely in the possibility of wrong decisions (misuse of nuclear weapons) on the part of information processors less capable and less powerful than the corresponding executors.

If we could establish an artificial intellect with powers equal to those of today's powerful executive organs, the danger of thermonuclear wars would be substantially decreased. (See also [1], where the author has written earlier on this topic.) In this case a new prospect arises—the expansion of artificial intelligence as one means of guaranteeing world security. Today measures are taken to hinder the expansion of powerful computers (embargoes on their export). Tomorrow, as soon as powerful programs are developed in important areas of control theory, it will be necessary not only to remove all restrictions but also to stimulate the expansion of powerful computing technology throughout the world.

A partial removal of the ban on the export of powerful computers is necessary even today, since they are needed for scientific research in the field of artificial intelligence.

These observations are based on the difference in principle between the spread of nuclear energy and the spread of artificial intelligence. Both can serve either peaceful or warlike goals. But however successful the expansion of nuclear energy for peaceful purposes, it does not decrease the danger that it will be used for military ends. Contrarily, the more widely artificial intelligence is applied in peaceful life, the more secure will people feel, since they will be wiser in the sense that they find better optimal variations.

CHAPTER 2
Methods of Limiting the Search Tree

Truncation

When the search tree is large it must be truncated. We may truncate near the root or further away. We can do the latter if the tree is narrow (Fig. 5). In a wide tree we must truncate near the initial position, i.e., the limiting depth of the variations is small.

The limiting depth has a strong influence on the precision of the solution. What depth do we want? This is a difficult question. We can more precisely forecast our prospects when we can more precisely assess the value of the variations in a search: the deeper the variation, the more precise the forecast.

The longer the optimal variation, the more time is needed to reach its end and the easier it is for us to correct our forecast if we must. For instance, suppose a chess master has evaluated a variation to a depth of 10 plies. After the first move in the variation, another ply can be added to the depth of the variation without changing the limiting depth. The variation has been extended by 10%; we are still relatively far from the region where variations do not exist (have not been computed) and where anything unexpected may occur.

If the computer evaluates variations generally to a depth of three plies, a single ply shortens its length by 33%, and we are much closer to the unknown future; the probability of error in our calculations is far higher than in the former case.

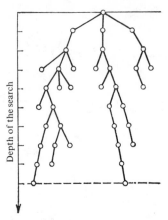

Figure 5   A deep and narrow pruned search tree.

The Goal of an Inexact Game

As soon as the search tree is truncated, the exact goal of the game loses all meaning. It is necessary to introduce an inexact goal in the truncated tree; else the game becomes aimless and cannot be strong. The goal of an inexact game permits the formation of a deep and narrow tree.

In chess the goal of the inexact game is to win material. A similar goal may be found in an arbitrary game that models a control system, and in an arbitrary inexact task. To attempt to solve an inexact problem without having formulated the goal of the corresponding inexact game is to waste time. This goal is the basis of a strong algorithm for the solution of an inexact problem, and the basis for development of a deep and narrow tree. We shall see later why this is so.

The goal of a game says what our aim is; only when we know this can we identify courses of action that cannot lead to our target, and exclude them from the tree. Knowing the goal lets us define the lines along which the search is to occur.

The Scoring Function

The goal lets us direct the search; the scoring function lets us evaluate and stop a variation. The goal lets us form a search tree; the scoring function lets us strike a balance.

The scoring function acts together with the goal of an inexact game and is therefore itself inexact. As distinct from the goal, which must be unique, the scoring function consists of two components: the first component allows us to evaluate the results obtained within the limits of the truncated tree; the second forecasts the possibility of reaching the goal beyond those limits.

The first component yields an exact answer (with the limited precision already determined) to the questions concerning the goals that have been reached; the second (the positional estimate) gives a preliminary answer to the question of what will happen later, when the boundary of the truncation has moved further away. Taken together, these two components determine the value of a concluded variation.

Breaking Off a Variation

The relationship between the value of the goal and the value of a variation allows us to decide whether to break off the variation before it reaches the limiting depth. We repeat that this is possible only when the algorithm contains both the goal of the play and the scoring function. When a part of the variation is cut off before the limit is reached, the tree shrinks (Fig. 5). As we have noted, all existing chess programs lack the goal of the inexact game, and therefore there are no variations shorter than the limiting depth.

The Pruning of Branches

We make no use of the well-known branch-and-bound method (the α-β-cutoff) in its customary form. Instead, we apply a simple method intimately bound up with the presence of an inexact goal for the game. If, during an ascent along a variation, the value of the current optimal variation (COV) at a higher node is not less than the sum of the current value (which was obtained at the given node during the descent) and the value of the goal (the target) of the same color as the node, there is no point in continuing along this branch, since such a branch can never generate a COV with a higher value.

The Horizon

The horizon method for bounding the search tree was introduced by the author in 1968 [1]. It is applied by humans in their everyday activities.

Suppose a pedestrian is crossing a road—he looks both right and left for automobiles. If an automobile is far away, i.e., beyond his "horizon" he may cross the highway; if it is within 50 meters he waits until it has passed.

Suppose an economist is taking part in planning a new factory. To guarantee the labor force, he must determine the numbers and composition of the population within some given radius from the factory. This is the economist's horizon.

Suppose an expedition is moving on sledges toward the North Pole and has to reconnoiter the ice conditions. Say the reconnaissance is made to a distance of one kilometer, but not all the way to the ultimate destination—the North Pole. This one-kilometer limit marks the horizon.

Use of the horizon is a rough but indispensable method for limiting the task.

We must not in any way confuse the horizon with the limiting depth of a variation (the distance to the cutoff point of the tree). The two concepts are quite different. Thus, the scouts exploring the ice may decide to go further after having surveyed the first kilometer, but each successive survey is limited to one kilometer.

The limiting depth of a variation always exceeds the horizon. Therefore the truncation of a search tree is always a milder process for limiting the task than the imposition of a horizon. (This refers only to the case when we are forming a deep and narrow tree.)

The formal definition of the horizon is as follows: The horizon is a limit to the amount of time that may be expended in reaching the goal. Therefore all potential goals which would consume more time in the reaching than is allowed by the horizon are to be considered as out of sight. The program solving the problem ignores them.

Two Trees: The Mathematical Model (MM)

The canonical solution of an inexact problem begins with the construction of a search tree. But if the problem is to be solved with a deep and narrow tree, the work must begin differently.

First we must find a goal within the limits of the accepted horizon, and only after the goal is known, and so are both the target we are aiming for and the direction in which to start the search, can we begin to build the tree. While constructing the tree within the limits of the horizon, we may find new goals, and then new directions become known for further formation of the tree.

Thus, in addition to the search tree, we form a tree of purposeful actions. The two trees are interrelated and closely interwoven. But we must remember that everything begins with the formation of the tree of purposeful actions—this lies at the base of all bases. The search tree can be grown in directions indicated by the action tree, although the latter continues to grow as the search tree grows. This makes for great economy in the use of resources, since the search tree is goal-directed and the action tree is formed within the limits that are necessary for the growth of the search tree—neither greater nor less.

The action tree may be called the mathematical model (MM) of the problem (task or game). An inexact problem can be solved only within the limits of a developed mathematical model.

The concept of the search tree is common to all types of inexact problems. The action tree is specific to each concrete inexact problem. In chess, the MM is a tree of trajectories consisting of purposeful moves of the pieces.

The search tree grows as it is formed; the MM should be kept bounded and as small as possible. (It is needed only for directing the growth of the search tree at any given node.) The smaller the MM, the more quickly we can inspect it and adopt the corresponding decision for controlling the search. Therefore the portion of the MM that has already played its role and is no longer needed should be erased from the computer's memory. A portion that is not needed at a given node of the search tree, but may be needed later, should also be cut out of the action at the current node but stored in memory. The limited active MM aids the formation of the search tree just as the patch of light thrown by the headlights of an automobile helps the driver to choose his path.

The Stratification of the System

We have already seen that the stratification of a control system contributes to the formation of a deep and narrow tree, because the search at any level is broken off as soon as the local goal is either attained in some way or found to be unattainable. We have also seen, and must not forget, that stratification leads to an increase in the resources that can be devoted to the solution of the problem, whenever there are control centers at the several levels that can solve the problem in parallel. This in turn contributes to the deepening of the search tree.

Another advantage of the multi-level system over the one-level system is worth noting. The plurality of goals allows us to avoid the inclusion of one or another level if the corresponding targets (goals) are such that their inclusion in the search would obviously lead to no change in the optimal variation (forecast). Moreover, using the value of the goal and the probability of attaining it, we can establish a priority list for the inclusion of the given level in the search; this also contributes to the rational formation of the search tree.

Thus the cutting off of various levels, or their ejection from the search, and the possible increase in resources that is characteristic of multi-level systems contribute to a deeper solution, because of the greater search limit and the more distant truncation of the tree.

Three General Limitation Principles

We have established the methods for limiting the search tree and for forming a deep and narrow tree (truncating, the goal of the game, scoring of a variation, the horizon, the mathematical model, and stratification). But without general principles governing the application of these methods, a successful deep and narrow tree cannot be obtained. What are these principles?

1 The Principle of Expectation. (Better called the principle of non-expectation, but we will leave the nomenclature unchanged, since the term "expectation" was established in 1968.) *As long as there exists a possibility for improvement, that possibility is included in the* MM *or in the search tree*.

The essence of the theory here is not only that these possible improvements take part in the solution of the problem, but also that possibilities giving no expectation of improvement are excluded from the MM and from the search tree. This is an extremely important step for the limitation of the problem.

This is how humans act. For example, a graduate of a music school who has become deaf does not persist in taking the entrance examinations for a music conservatory, even though he had been successful until he lost his hearing. A chess player does not persist in contemplating the capture of a Pawn (for lack, say, of any other goal) if his Rook perishes in the attempt to reach the Pawn.

2 The Principle of Maximum Gain. *A new possibility (in the* MM *or in the search tree) is to be considered only if it offers an expectation of gaining more than the possibilities already considered.*

For instance, a chess master may be able to win a Knight for a Pawn; now he contemplates another possibility, which will result in winning a Pawn. The new opportunity is disregarded, but another is found—an attack on his opponent's King. The principle of maximum gain says that this latter potential must certainly be kept under consideration.

Again, our graduate must decide where he should apply for entrance —University A or University B. He prefers A, but the competition for entrance to B is substantially less. He must ask himself whether he should send his credentials to A. But if the competition at both were the same, the question would not even arise.

This principle, too, leads to a reduction in the number of possibilities under consideration.

3 The Principle of Timeliness. *Only those possibilities should be considered in which the actors have time to take part in the game.*

There is no point in involving a piece in the control of some field on the chess board if that piece has time only to begin to control the field before an enemy piece slips in.

The commander of an anti-aircraft defense zone will not dispatch a plane on an interceptor mission if it would in any case be too late. Instead he sends an interceptor from an airfield that allows the pilot time to attack the target.

The principle of timeliness (or should it be untimeliness?) is also a powerful method for eliminating possibilities that are senseless or deficient in sense.

These three principles are interrelated; that is, in each a possibility is excluded if there is no hope of its timely participation in the game or if it obviously offers no hope of greater success in reaching the goal.

Improving the Results of a Search

During the development of a MM, a question inevitably arises: Is a given concrete extension worthwhile? Is there hope of greater gain? Of a better result of the search? In other words, will the extension of the MM yield a new optimal variation with a higher value? The decision is to be made by the use of our three principles of expectation, maximum gain, and timeliness. The questions cannot be answered in general, but only in concrete situations for specific problems; nevertheless some general recommendations may be made.

Mathematicians investigate the search tree variation by variation, using a minimax procedure. Then, having found a variation and stowed it in the computer memory, they find another. The variations are compared via the scoring function; the one with the highest value is retained in memory, and the rest are discarded. The process is then repeated. It is not difficult to see that only two variations are present simultaneously in the computer memory, and one of these is chosen as the current optimal variation (COV).

Since the conventional search for the optimal variation carries through to the solution of the problem by including all variations in the truncated tree, this method cannot be used for a solution based on a deep and narrow tree. Such a tree can be formed only by restricting the MM, not allowing it to grow, which we do by applying our three principles. We cannot apply them if we keep only two variations in view at the same time.

The essence of the matter is that when we are forming a subtree below a given node and must decide whether to extend the MM, we can do so only after investigating the variations in the subtree. Only by collecting concrete information about these variations (including the COV) can we answer the question: Can we expect to find a new COV with higher value?

We must remember the subtree, since we must collect information about the variations in it. Here is one of the distinctions between the method of solution by forming a deep and narrow search tree and the method that includes all possibilities in the search tree.

It is well known that a chess master does not remember two variations only but, rather, remembers the search tree. He keeps in mind the whole tree as finally formed, which is the basis for his decision.

In summary, we have considered methods for restricting the search tree. These may be divided into two types: those with respect to time and those with respect to material (the goal of the game). The truncation of the tree occurs with respect to time, the horizon sets a time limit for reaching the goal, the principle of timeliness invokes time as its essence. The principle of maximum gain, the improvement of the results of the search, the breaking off of a variation, and the scoring function are related to gain of material (the goal of the game). We may add that a positional estimate, although perhaps not immediately connected with material, nevertheless depends on it, as we shall see later.

CHAPTER 3

The Search for a Solution and Historical Experience

When we meet an original situation and wish to use a decision-making program, we apply some of our historical experience indirectly, in the development of our algorithm and in the structure of the program. We form a deep and narrow tree of suitable size, expending substantial resources in time and memory. Usually, however, we also have direct historical experience, accumulated over the centuries and specifically related to the concrete problem that is to be solved. If we apply this experience, we can save our resources and obtain a faster and deeper solution. A specialist solving a problem always uses his knowledge and experience if the situation permits.

The Search for a Solution by Association

A human, who while solving an inexact problem finds that the situation reminds him of one that people have met before, applies historical experience to the solution.

What does "reminding" him of a situation mean? His current situation is obviously not exactly the same as the one that he or others have met before —the resemblance is only partial. But it provides a key to a simpler method for finding the solution than the one he would have used in the original situation, namely, he has a ready-made method for developing a portion of the MM and, correspondingly, a portion of the search tree. In chess, we see that we must form just that portion of the MM in which the same pieces that we are now concerned with take part. In the past, these pieces have been moved in certain ways, and we must now test whether the variations that have been tried before have been successful. If they have been, the search for a decision in the given subtree is ended; if they have not, the remaining portions of the MM and of the subtree that was to be formed are not needed.

The associative method suggests to the program the line it should follow in searching variations; in other words, this method defines the priorities for the game at a given level of control. Since this amounts to the formation of a part of the MM and a partial formation of the search tree, it economizes on resources. The savings in time and memory can be applied to the continuation of the search for the solution in the original situation; the solutions that are found will be deeper.

The Handbook Method of Searching

Here everything is simple: If a situation arises that is identical to one known in history, the solution has been found, and no further search is needed. Since the situation has already been met, its value is known; consequently the value of the variation is also known, and the variation may be broken off. In this case, we obtain the value of the variation, from the moment when the situations coincide, without further development of the MM and the search tree. This is a highly economical method.

The trick here is to try to find an advantageous coincidence of situations. This leads to a directed development of the MM and the search tree, but once the coincidence is reached and the value is known, the variation is ended.

So, we have considered three types of search.

In the first type, in an original situation, when history is of no help, we must construct the MM and the corresponding subtree by the standard methods.

In the second type, when the situation has been partially encountered in the past, we know the direction of further development of part of the MM and subtree. If this is successful and leads to an essential coincidence of the situations, not merely an external resemblance, further development of the MM is unnecessary. There is a saving of resources. This method requires a partial formation of the MM and the corresponding subtree.

Finally, in the third type, the given situation coincides completely with one found in history (for this method we may need to make a directed search). The value of the variation is then already known without further expenditure of resources. This is the handbook method.

A human specialist in some given area applies all three methods to the search for a solution when he is dealing with an inexact problem in his own specialty. He also uses his special training. A computer should act in the same way; it should be provided with historical experience, special knowledge, and training in the use of all three methods.

A chess program should be provided with special knowledge of the openings, the middle game, and the endgame in such a form that the knowledge can be used for either the associative or the handbook method of search.

CHAPTER 4

An Example of the Solution of an Inexact Problem (Chess)

We have set forth the general postulates on which the solution of an inexact problem should be based. We shall illustrate their application by a concrete example—the development of a chess-playing program for an electronic computer. Chess, as we have already noted, is a typical inexact game—a model of a multi-level control system with a single information processor.

When the algorithm for the chess program was developed, the task of modelling the thought processes of the chess master was posed. We assumed that mankind, in the many centuries of chess history, had adapted rationally to this inexact task, and therefore decided, without worrying the question, to translate human experience for the machine.

A master uses two methods of play: (1) an algorithm for finding a solution in an original situation (position) and (2) an algorithm for finding moves in positions that have been met, in part or in whole, earlier in the experience of chess masters (and here he himself uses his own specialized knowledge of the game of chess). The computer should be provided with both methods of play, and must therefore be given not only a search algorithm but also a knowledge of the openings, middle games, and endgames.

Almost all chess-playing programs have a library of the openings; this is a simple matter. They have been given no such libraries of the middle and endgames. We shall see later why programs based on the inclusion of all possible moves in the truncated tree are organically deprived of the ability to make use of middle and endgame libraries.

A. The Search for a Move in an Original Situation

The Truncated Search Tree

Chess is a complex game, though soluble by enumeration of its moves. In theory this enumeration is finite, and therefore chess is a finite game. In practice, however, the search tree for the moves is so immeasurably large that no chess player can manage it without truncation; when the tree is truncated, chess becomes, as we have seen, an inexact game.

We have already noted that the tree widens extremely rapidly if we solve the problem by including all moves in it (see Fig. 1) and our task becomes hopeless. If, on the other hand, we solve the problem by using a deep and narrow tree, as a chess player does, there is hope for a successful solution of the problem of searching for a move.

The depth of the truncation (the limiting depth of a variation) is the first constraint on the problem. By changing the limiting depth, the chess player controls the number of moves included in the search tree. With a deep search tree, a change in the limiting depth is a delicate method for restricting the search.

We took the formation of a deep and narrow tree as a basic requirement in the development of our program. Our algorithm as proposed was to be like that of a chess master; the tree as formed must therefore be deep and narrow, and contain only a small number of moves.

Are we ready, with this, to begin work?

The Goal of the Inexact Game in Chess

According to the general theory developed above, the work must begin with the establishment of the goal of the inexact game, and we shall act accordingly.

Every chess master knows his aim when he sits down at the board—to win material. The amateur is horrified by this assertion—and his actions are in vain! All other factors, including the positional, which the master takes into account in his reasonings (or, more exactly, in applying a minimax procedure to the truncated search tree of moves), serve only this prosaic aim.

This goal is a model which appears in quite distinct specific forms, but as a type it is invariant.

In establishing the goal of the inexact game, we simultaneously obtain our first criterion, allowing us, at the beginning of the development of the search tree, to define senseless moves. When we know nothing about the position and are only beginning to study it, we exclude from the search all moves which do not lead to a possible gain in material, to an attack by the pieces of one color on the pieces of the opposite color. Instead of using the primitive procedure of including all moves in the search, everything here begins by defining moves that make no sense. This must be done, there being no reasonable goal for the game as yet.

We have said earlier that chess, as played by a master, is a model for a three-level control system. For example, the first level consists of an attacking piece with its attack trajectory against a piece of opposite color; this level has its own specific goal—the annihilation of the target piece, the winning of material at a price no higher than the value of the target. The second level refers to the field of action, for instance the ensemble of pieces taking part by hindering or supporting the action of the attacking piece; this level also has, in a very different form, the goal of winning material. And finally the third level, the mathematical model of the position, represents an ensemble of fields. For the third level there are characteristically many elementary goals, since in the MM various pieces of different colors are on the attack, but the attainment of all these goals should lead to a gain in material.

We note that in chess the goal of the inexact game is of the same type as the goal of the exact game. We must not forget that in chess the King has an infinitely large value (of course a chess program cannot conveniently attach an infinite value, and so customarily attaches the value 200 to the King). But whether the value is taken to be infinite or 200, the value of the King is a material value, so that the goals of the exact and inexact games are of the same type.

The Scoring Function: Two Components

We have already noted that the goal of a game tells us what we are aiming at, and the scoring function is used to evaluate both the results already obtained and those that may be obtained.

What sort of scoring function is required in chess? We already know that it must have two components: The first allows us to judge how successfully we may attain the goal of the inexact game (a gain in material) within the limits of the minimax procedure on the truncated search tree. In simple terms, we add up the values of the pieces won by the two sides, in each variation within these limits; this component is clear and simple. The second component is more complex; it has to forecast the gain of material in the as-yet-unknown portion of the tree that lies beyond the cut-line. This amounts to a positional estimate.

The scoring function, according to Shannon, appears to consist of many parts, but in reality it consists of only two. The first deals with material; the second deals with position and is pieced together from very distinct positional factors, the value of each being derived by some kind of averaging process.

Such a method of deriving a positional estimate in chess is wrong. A positional factor that yields a positive result in one situation may yield a negative result in another. For instance, are doubled pawns good or bad? The answer depends on the situation. Sometimes doubled pawns are a suitable target for an attack, since one of them cannot support the other. But sometimes doubled pawns contribute to the control of squares through which important communication lines (trajectories of pieces) pass, and then the doubling is extremely useful. The same thing may be said of other factors entering into the positional component of Shannon's prototypical scoring function.

In the chess program under consideration, a quite different decision was made about the positional component of the scoring function, basing it on the control of those squares making up the trajectories that enter the MM. The side controlling the larger number of squares has a positional preponderance. Later we shall examine the question of the positional estimate in more detail. We note here, however, that the control of a square is determined by the result of exchanges that take place on it, i.e., by material relationships, and the positional estimate is needed only to forecast the winning of that material.

Breaking Off a Variation and Suspension of Play

The goal of the inexact game (material) lets a variation be terminated before it has reached the limiting length. Let us denote the value of the goal by m_g and of the current estimate as we pass along the variation by m_c. If we consider a single field, we continue the variation if $|m_g| > |m_c|$; otherwise we break it off.

This inequality expresses the principles of expectation and maximum gain: as long as there is hope of reaching the goal, as long as less than the value of the goal is lost or won, the variation continues. When, however, the numerical value of the lost or gained material equals or exceeds the value of the goal, play is ended and the variation is broken off.

The essence of the principle is this, that in chess, as in other tasks, material may be sacrificed only if there is an expectation of winning material of greater value or of avoiding the loss of material having a value greater than the material sacrificed. When material is sacrificed, both the principle of maximum gain and the principle of expectation are expressed.

A variation may be broken off for other reasons, also connected with the possibility of reaching the goal or of being unable to reach it, e.g., when the target or the attacking piece vanishes—and here we invoke the principles of expectation and timeliness.

In general, however, a variation is ended when play is suspended at the highest level—the ensemble of all other levels. In the lower levels, play may be broken off while the variation is still being continued. Play is suspended at these levels when the corresponding goal is reached or is found to be unreachable, or when there is not enough time to reach the scene of battle. We shall return to this topic after formalizing the multi-level system for chess. In the suspension of play and stopping of variations, we see the difference between the algorithm under discussion and other chess-playing algorithms. These two factors contribute to a decrease in the number of moves in the search tree.

The pruning of branches is also founded on the principles of expectation and maximum gain. If the sum of the goals at a node is less than the difference between the current value and that of the current optimal variation, there is no hope of increasing the latter by continuing from the given node, and a further search below the node must be stopped. We must remember that the sum of the goals at a node may change its value when an attack field is brought into the play; we must take this into account when pruning a branch.

It follows from what we have said above that in our algorithm the pruning of branches has nothing whatever to do with the branch-and-bound procedure; it is connected only with the goal of the play—when it is impossible to improve the results of pursuing the goal, nothing is to be gained by constructing the branch.

The Horizon

There may exist an uncalculable set of attacks by a piece of one color on opposing pieces. The goal of an inexact game gives a meaning to the action of pieces but cannot by itself limit their action. We must introduce a limiting method based on the already mentioned horizon principle. In our chess program we introduce a limit on the amount of time available for pieces to move along their attack trajectories. The horizon method expresses the principle of timeliness: an attack that cannot be completed within the horizon is excluded from consideration.

We know that in chess time is measured in half-moves, or plies. If we adopt a horizon H_L equal to an even number of plies, the limiting number of moves by the attacking piece along its attack trajectory will be $H_L/2$ (this does not apply to a denial trajectory).

The limiting horizon H_L is a harsh bound, since a change in the horizon leads to a sharp change in the number of moves in the search tree. When there are many pieces on the board, the play takes on an open character and the horizon normally consists of four plies. In the endgame, when there are few pieces, and especially if long-range pieces are even partially removed from the board, the number of attacks that can be mounted within a horizon $H_L = 4$ is sharply reduced and the horizon must be extended in order to find good solutions in the search for moves. In particular, the horizon in pawn endings is large, extending perhaps to 10 plies.

Chess as a Three-Level System

There can be no doubt that a strong chess player begins all of his calculations with the attack, within the horizon. There can also be no doubt that a strong player sees, not the whole board and all the pieces, but only those which appear in his calculations and move through certain squares on the board. The ensemble of these pieces and their trajectories make up the mathematical model (MM), the tree of trajectories that underlies the search tree of moves. So there must exist two levels: pieces with their trajectories, and the MM.

But there is an intermediate level which makes chess a three-level system. Everything begins at the first level—pieces and their trajectories, lying within the limits of the horizon H_L. If a piece has its individual goal, it is an attacking piece. If a piece in its trajectory, within the horizon, takes part in play connected only with the struggle of another piece to attain its individual goal, then it is not an attacking piece. *Pieces whose trajectories are determined by the horizon are called stem pieces.*

An attacking piece that is also a stem piece attempts to attain a well-defined goal. A stem piece that is not an attacking piece tries only to change the result of the assault on an attacking piece's goal.

Stem pieces do not act alone. Each has a crew of pieces of its own color who support it; there is also a hostile crew of pieces of the opposite color who hinder it. The ensemble of stem pieces and the two sets of commands of pieces of both colors form a field of play—and this is the second (intermediate) level of the control system. Since the stem pieces are of two types (attacking and attack-dependent), the fields must also be of the same two types.

Thus, the second level may consist of an attack field or dependent fields. The latter have been called control, blockade, deblockade, retreat (when the target disappears), and positional (when the sign of the exchange is altered in a controlled field). The second level (field) is always formed by one side to improve the results of a search (improve the value of the COV). If it is an attack field, new targets may appear (in this way, a second level of attack type may influence the value of the COV).

The First Level—A Piece with a Trajectory

The first level may be of one of two types. A piece and its trajectory may belong to a stem lying at the base of a (second level) field of play; the limiting length of the trajectory is constrained by the horizon H_L. A first level of this type is formed only when the second level (field) being formed can change the value of the COV in a favorable direction.

The situation is more complex for first level entities of the second type—pieces with non-stem trajectories, i.e., those in the supporting or opposing crews, the so-called denial pieces acting in a denial trajectory. In keeping with the principle of timeliness, the horizon H_x for such pieces is not fixed, but variable. The denial piece is in the final analysis connected with the battle for an a- or b-square of the stem trajectories of the field (on an a-square, a piece halts in its trajectory; it moves over a b-square without stopping). The larger the number of this a-square, i.e., the further it lies from the initial a_0-square of the stem trajectory, the more time a denial piece has for its arrival and participation in the battle, and the larger the value of the variable horizon H_x. Interestingly enough, H_x for a control denial trajectory may exceed H_L by a half-move; this may happen in connection with the control of the a_f-square (the last) of a stem trajectory, where the controlling side has the move and the stem trajectory is of limiting length. The complexity in the formation of denial trajectories arises from the fact that the time limit H_x is concerned not only with the formation of the first denial trajectory, which is immediately connected with an a-square of a stem trajectory, but also with all subsequent denial trajectories, from the first to the higher ones. A high-order denial is connected with an a-square of a denial trajectory of order less by one, and in the final analysis, with a given a-square in a stem trajectory.

The principal peculiarity in the formation of a denial trajectory, as opposed to a stem trajectory, is that if such trajectories can exist within the limits H_x, they are formed unreservedly. The only principle to be observed is that of timeliness.

We shall show later that the inclusion and exclusion of already formed denial trajectories in the MM occur in accordance with the principle of timeliness.

In forming stem trajectories, both the principle of timeliness (H_L) and the principle of maximum gain (improvement of the value of the COV) apply; in the formation of denial trajectories, on the other hand, only the principle of timeliness applies. This is understandable: the construction of a field can proceed as long as we have no information as to its unconditional uselessness; thus the denial trajectories must be formed.

The principle of expectation applies to both stem and denial trajectories —it is always applied.

We must remember that not all pieces with their trajectories form first-level control systems, but stem pieces with their attack trajectories certainly do, since they have their individual goals, as control systems at all levels must.

When a controlling piece acts in a denial trajectory, the situation is more complex. A denial trajectory can exist only in a field of play—this means it already belongs to a second-level control system. But at the same time, the controlling piece has its own target: it lies in ambush, to strike its own victim. Once such a goal exists, of course, a piece acting in a control trajectory may be moved to the first control system level.

A piece acting in a retreat trajectory should be taken as belonging to the first level, since it has its individual goal—to rescue itself.

A blockading or deblockading piece, on the other hand, cannot belong to the first level, since it has no individual goal. In a blockade (if we are not considering the blockade of a Pawn), the piece is prepared to sacrifice itself, i.e., it acts as one with its colleagues (of the same color). Such a piece enters only into the second-level control systems. A deblockading piece acts out of the same general interests and belongs to a second-level system.

The Second Level—A Field of Play

An ensemble of stem and denial pieces forms a second-level system, a field of play. There may exist attack fields, which include concrete goals (targets) in the play, and dependent fields that do not have their own targets. Both types are formed only when it is expected that their inclusion in the play will lead to an increase in the value of the COV.

Fields are formed, in keeping with the three fundamental principles, when there is hope that a stem piece will succeed in making a gain, i.e., when the possibility of increasing the value of the COV is not excluded.

The variable horizon H_x defines a strongly determined structure of the field: the field includes those denial pieces that can take part in the play within the time limit H_x. All denial trajectories satisfying this limit must unconditionally be included in the field. (The possibility of obtaining all denial trajectories has been verified by experimental studies of chess problems.) Figure 6 displays three fields of play: an attack field Z_0 and two connected fields, Z_1 (control) and Z_2 (retreat).

The Third Level—The Ensemble of All Fields (MM)

We have already discussed the conditions for forming an ensemble of fields. The mathematical model includes only the fields in which there is hope for increasing the value of the COV. Thus the volume of the MM depends in particular on the first move in the initial position; given a successful choice of this move, further inclusion of fields may be minimized or eliminated entirely, leading to a significant saving of resources. For this purpose we need a system of priorities.

With this preliminary introduction to the third-level control system—the game of chess itself—we now proceed to a careful analysis.

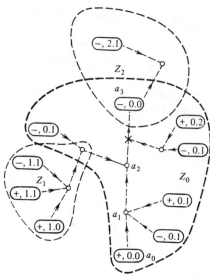

Figure 6 Fields of play. 0, 1, and 2 are, respectively the stem trajectory, first denial trajectory, and second denial trajectory. The symbol "+" denotes color of the stem piece of the field; the symbol "−" denotes the color opposing the stem piece; and 0, 1 denote the pieces of first denial in the zero field belonging to the attack.

The Search Tree and the Minimax Procedure

The method for constructing a search tree is well known. A variation in the search is continued from its initial position to its end (the rule for ending it is irrelevant here—whether it be logical or by limiting length). Then the ascent of the variation begins. At the first node up from the bottom, we must test the possibility of forming new branches of the tree. If none exists we continue to climb; if there is a possibility, we form a new branch and go down the tree to the end of the new variation, and so on.

Constructing the tree is not enough; we must also score it, i.e., find the optimal variation. This is done by applying a minimax procedure. The scoring function is applied at the end of every branch, and in the ascent up the variation, the resulting value is attached to the nearest higher node. When at a given node the values of all variations joining the tree below it have been collected, the player having the move at the next lower node (he has the right to choose the variation according to its value) chooses the one that gives him maximum value, which is minimal for his opponent. The variation with this value is the COV at the given node.

In the initial position, the highest node on the tree, the COV becomes the optimal variation, which determines the choice of the first move. (If the values of several different moves are equal, the program, like a human player, throws a die.) The process we have described constitutes the minimax procedure on a search tree. When the tree is very large and when the memory available for storing it is too small, the minimax procedure holds in memory only two variations attached to the given node. The values of these two are compared; the one with least value is erased, and the descent along a new variation is begun. After the end has been reached and the return climb has reached the given node, the value of the new variation is attached and compared with the one residing in memory, one or the other is selected, and so on.

The comparison of two values appears to be the method used by all programs actually playing. We note that comparison is necessary when the search tree is very large, but leads to feeble results in the search for a move. A human uses a different method: he forms a small tree and keeps all of it in mind. This allows him to find a good move in the initial position.

When we can keep in memory the subordinate search tree lying below the node at which we are constructing the COV, we can form variations attached to the given node and having favorable values. When the player trying to improve his COV expects to find one or more variations having a higher value than that of the COV, he forms a supplementary branch of the subtree. The final determination of the COV and the lifting of its value to the successor node is completed only when there is no further expectation of changing its value by the minimax procedure.

Next we consider the technical construction of the mathematical model.

The Technique for Defining Trajectories

The shortest path for the movement of a piece from one square to another is determined with the aid of a conceptual board of 15×15 squares. It is assumed that any other path is no more than the concatenation of two shortest paths. For long-range pieces (Queen, Rook, Bishop) it is also assumed that the shortest path (trajectory) consists of two moves (three a-squares). These assumptions greatly simplify the task of determining trajectories and probably do not significantly worsen the choice of a move.

Chess is noted for the wide variety of ways in which the several pieces move. With the help of the 15×15 board, however, all the various trajectories can be obtained; one might suggest that chess masters use such a conceptual board in finding trajectories.

The ability to construct a trajectory allows us to construct the first-level control system—pieces moving in their trajectories—with relative ease.

We must also note that when the displacement of a piece from one square to another is defined, a unique trajectory exists only when the interval between the squares can be covered in a single move. For displacements requiring more than one move, there exists a sheaf of trajectories, since as a rule a piece may move from one square to another in a variety of ways. Thus, essentially, when we speak of determining a trajectory, we mean determining a corresponding sheaf.

The Technique for Forming Fields

The construction of the second level of play in chess is a subtle and delicate affair. There are two types of fields: (1) when, for instance, a stem piece in an a_0-square tries to move to a final a_f-square along one of the possible trajectories in the sheaf of stem trajectories, and (2) when a piece attempts to abandon an a_0-square on which it stands and move in a single step to one of the nearest a_f-squares. In the first case the final square is unique; in the second, there is a set of final squares. Fields of attack, control, and blockade belong to the first type; fields of retreat and deblockade belong to the second.

Retreat and deblockade fields have an elementary structure. In the general case, attack, control, and blockade fields have a rather complex structure.

As we noted earlier, the principle of timeliness underlies the construction of fields of type (1). We shall provisionally assume that among the pieces of type $(+)$ [these are pieces of the same color as the stem piece; those of opposite color are designated by $(-)$], only the stem piece itself can move. Other $(+)$ pieces may move only to capture $(-)$ pieces in a single move; in other words, they take part in the play only if they are already in ambush. Pieces of type $(-)$ are included in the field and are active only if they have time to take part in the play; the determination is formalized with the aid of the variable horizon H_x.

The more a-squares there are in a stem trajectory, i.e., the more half-moves a stem piece must make to reach a given a-square in a stem trajectory, the larger the horizon H_x for that square and the greater the number of half-moves available to $(-)$ pieces connected to the fight for it. The machinery for determining H_x and apportioning the corresponding number of half-moves among the denial trajectories of $(-)$ pieces was developed by B. M. Stilman.

Stilman also suggested a connection between the construction of fields and the construction of the search tree. This is extremely useful, since under these circumstances a field is generally constructed not completely but only partially, only as far as necessary. If the search of a partially formed field must be cut off, so must the further construction of the field.

The problem of defining denial trajectories for the $(-)$ pieces is complex. Those within the horizon H_x must be defined completely, without exception, since until the results of the search are known, we cannot say which of the denial trajectories are useless.

They are determined by a two-part method. For instance, if during the play in a field some piece moves to a new a-square, we test to see whether some other piece (or pieces) can attack the moving piece, within the time horizon H_x. If the answer is yes, we determine the denial trajectory. This method does not generate all denial trajectories, since in counting the pieces that are to act in denial trajectories as yet undefined, a possible trajectory might leave from the a_0-square and be overlooked.

To define all the omitted denial trajectories, a second method was introduced. Its essence is a new procedure for ascending the variation (as along a track already traversed) in order to find missing denial trajectories that may have some a-square of the search at the given node as their final square. Then the pieces leaving the initial squares of the possible denial trajectories will return to these squares and, since the horizon H_x has already been determined, the problem of determining all denial trajectories can be solved. As soon as a new denial trajectory is included in the play, the machinery for the pseudo-search is invoked; we return along the variation to the node at which the new denial trajectory appeared, and the search continues.

In the return along the variation (in the pseudo-search procedure), we must not erase the subtree rooted at the given node where the denial trajectory was included; only some branches are deleted. (See Fig. 7).

There is a possibility of deliverance from this expensive operation of erasing a subtree, if contrary to the usual procedure for forming a tree, we admit repeated moves at a node. We shall discuss this later. Thus, when a new denial trajectory appears, we start the pseudo-search, backtrack along the variation, erase a branch of the subtree (for which the value of the variation has not been determined), and consequently use up a certain amount of our resources.

As can be seen from Fig. 7, we erase only one unevaluated branch (nk in the figure); all the remaining variations have been given their final values and are to be conserved. The denial trajectories are unconditionally included in the play, but, as we shall see later, with assigned priorities. We may add that in the earlier stages of our program development, the problem of erasing branches in the pseudo-search was incorrectly solved—the entire

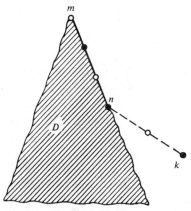

Figure 7   The essence of the pseudo-search procedure: D—the subtree already formed; k—the node at which a denial trajectory has been found; n—the node at which the branch from n to k must be erased (the dashed line); m—the node at which the search is supported, i.e., the node for inclusion of the denial trajectory.

subtree below *m* was erased; this was necessary because the repetition of moves at a node had not been introduced into the program.

In 1960, in a lecture at Humboldt University in Berlin, the author remarked intuitively that a chess master proceeds by a method of sequential approximation. The pseudo-search is an obvious example.

Having become acquainted with the general features of the technique for forming fields, we may now go further.

The Positional Component of the Scoring Function

The positional estimate should not be a general-purpose affair; it should be specific to each given situation. A general-purpose estimate might, for instance, be applicable to 67% of the positions encountered, and wrong in 33%; the current position might fall in the latter group. We must give Capablanca due recognition in this respect; in a polemic with Znosko-Borovsky, author of *The Theory of the Middle Game in Chess*, Capablanca pointed out that the basis for a positional estimate is the control of fields. He also condemned such abstract concepts as "space" and "time".

Today, when the machinery for formalizing the methods of play used by masters has been basically created, we can formalize the notion of "control of fields".

Control of fields does not mean control of the whole board, but control of only those fields that may be used in the impending play. Therefore one must strive for control of the field consisting of those trajectories in which the pieces can move, but have not moved yet.

At the node in the search tree where we find ourselves at a given moment, we must unravel all those sheaves of trajectories which have not yet been developed and determine which player has control of the majority of the fields consisting of the trajectories not yet used in the play. This allows us to forecast the result of the play—the result of a search which, in particular, had to be renounced at the terminal nodes of the variations for lack of resources.

We may assume that a so-called positional sacrifice should not exceed two units of material (this will be made more precise in an experiment). Therefore, if one player at a node where a variation is broken off has lost more than two units, it might appear that the positional estimate cannot change anything (the variation may of course have a negative value). This is wrong, since the influence of the positional value on the minimax remains.

We shall show later that the positional estimate allows us to solve the question of priorities for including pieces and fields in the search at a given node in the tree. Thus the positional estimate, with a development of the sheaf of trajectories, should be produced at every node in the search tree. We may assert that the squares under control define the usable mobility and maneuverability of the pieces. Better maneuverability of pieces often also determines the positional superiority.

It was assumed that the control of squares involves only those pieces that lie at a distance of one move from the controlled square (and a blockading piece must lie on the square itself). The sign of the exchange of the pieces (at a square) that take part in the positional estimate also defines the control of the square as belonging to the side for whom the exchange is favorable.

To sum up: The positional estimate is computed at every node of the search tree. The procedure is substantially more complex than the procedure for computing a material score. All sheaves of trajectories included in the play but not yet used (in whole or in part) are taken into account in computing the positional value. We must emphasize a difference between material and positional estimates. A material estimate at a given node is invariant; it does not depend on how we arrived at this node, whether from above or from below. A positional estimate at a given node may vary, since additional fields may be included in either descent or ascent. When a new field, a new sheaf of trajectories, is added, the positional estimate may change. Thus it varies much in the same way as the sum of the values of the goals (targets).

Nevertheless, we note that the construction of the positional estimate is much simpler than that of fields, since here the horizon H_x is either 1 or 0 for both Black and White. The positional estimate at a given square in a trajectory is computed only up to, but excluding, the first square at which it ceases to be positive, and this determines the movement of a piece along its trajectory.

This all happens very logically: since the variation is carried no further, we see that we may get a first approximation on those trajectories on which the pieces have not yet had time to move.

Thus the basic factor in the positional estimate is proportional to the ratio K_w/K_b, where K_w and K_b are the numbers of squares controlled by White and Black, respectively.

Priorities for Including in the Search Those Pieces and Fields that Take Part in the Play

First of all, we must decide whether or not to include a given field in the play. If we do decide to include it (we shall see later how this is done) we must further decide the order in which the fields and pieces should be included in the search. This is a most important decision. The size of the search tree depends on the priorities we assign. If we choose an unfortunate priority order, the tree will be larger. A shortage of resources leads to a shortening of the limiting length of variations, that is, to a more shallow solution of the task.

The priority for inclusion must depend on the positional estimate, that is, on the control of squares. We need two more concepts, which we now introduce.

Vulnerable target. If all the squares in an attack trajectory and all the squares in the target's retreat trajectory are controlled by the attacking side, we shall say that the target is vulnerable. We set the value of the target equal to the smallest result of the exchanges on the squares in the trajectory, provided this result does not exceed the value of the target itself.

If instead of an attack trajectory, we consider any other one, the same concepts apply with one change: the value of the target will be set equal to the value of the target in the attack field with which our arbitrary trajectory is connected.

Strikable target. This is a vulnerable target on a stem trajectory of attack for which the trajectory is one move long, so that the target can be annihilated and the attacking side is to move. Under these conditions the target may be struck; that is, without making the capture, the result may be predicted exactly.

We establish the following priorities for including trajectories in the search (in fields): if the value of a vulnerable target is such that it makes sense to carry out play in the given field, then, first of all, the field should be played. If there exists another vulnerable target having the same value, play should be executed in the field where the goal is most quickly reached. Given different values of vulnerable targets, play should occur in the field containing the target of highest value.

The experience of the chess master shows that this is the best way to find a good solution with minimal size of the search tree.

The notion of the strikable target also contributes to the discovery of small trees. For instance, if White can strike a target of value such that the variation will be immediately broken off (or if this value is equal to the difference between White's and Black's strikable targets), the variation striking the target need not and should not be continued—and this clarifies everything.

A second example: If there are both strikable and vulnerable targets of the same color, we should play against the strikable; the vulnerable one will wait.

We may shrink the search tree by using the notions of strikable and vulnerable targets for non-stem trajectories as well. For instance, after we have unravelled a sheaf of trajectories, the following question arises: Which of the trajectories comes first? The answer is clear—the one in which all the squares are under control, and of these, the shortest. If we must decide which sheaf to play in first, we use the same method for decision.

The forking of trajectories also determines their priority for inclusion in the search. When they coincide over some portion of their length, a piece moving over the common (fork-handle) portion will move with greater speed when the common portion is longer. This also yields a priority for inclusion, but of less value than in the previous cases.

The positional estimate which we have computed—the ratio of the total numbers of squares controlled by White and Black in the trajectories (up to the first uncontrolled square)—tells us about the relative mobility and freedom of maneuver of the pieces, as we have seen earlier.

The positional estimate should depend on other factors as well. There is no doubt that it should depend on the ratio of the sums of the values of the vulnerable targets, which can forecast the degree of success to be obtained below the summit of the truncated tree. Time must also be taken into account, namely, the time that must be spent in moving the attacking pieces along the trajectories of vulnerability. The smaller the total time, the better. Therefore the ratio of these total times must also be taken into account. Probably it should enter the calculations only when White and Black have equal sums of the values of vulnerable targets, but this question can be decided only by experiment.

In summary, the three classical factors do enter the positional estimate: material, space, and time are all included, but they are not at all the same factors written about in chess manuals.

The Inclusion of a Field in the Play

As a rule, the question of including a field in the play should be decided by the use of the lowest possible node in the subtree. This means that a variation in the searching process is carried through to its end, and then, in the backtracking climb, the decision is made at the first node up from the end where the question arose.

Thus, as a rule, in the tree under the node where the question of including a field arises, all earlier questions about the inclusion of fields should have already been settled, so that below the node in question everything is clean.

In the process of searching moves within the horizon H_L, a set of new perceptions may arise, when a piece of one color sees a piece of the opposite color (that is, when the existence of a trajectory is established). This leads to the formation of a sheaf of stem trajectories, after which we must decide whether or not to include the field.

It should be excluded unless it offers hope of improving the value of the COV. We may formulate the following rule: Exclude a field if it satisfies one of the two following conditions: (1) the piece obviously cannot reach the a_f-square of its stem trajectory or (2) inclusion of the field obviously cannot change the value of the COV.

We test these conditions in the following way: We assume that the subtree under the node where the decision is to be made has already been formed. We must first convince ourselves that the field in question is not already included in our variation, which must therefore be investigated. To do this we introduce the "traffic signal" principle: a move relating to one field only is marked to indicate the fact and the field; if the field under investigation encounters such a sign, a metaphorical red light is turned on, but if there is no encounter, or if the move relates to several fields, the light is green.

Since we are discussing the improvement of the COV, let us be more precise about the fact that in determining whether or not a piece can arrive in time to act, we need not investigate all the variations in the subtree. If, for instance, we are studying the subtree under a White node and we solve the problem of including a White field, we must investigate only those variations that coincide with the COV at Black nodes and take into account only those trajectories that have appeared in these variations. (See Fig. 8.) It makes no sense to investigate other variations, since a favorable change of value in them cannot favorably change the minimax result. Let us assume that the investigation of this portion of the subtree has been carried out with a positive result—the piece can indeed arrive in sufficient time at the a_f-square. Then we must answer the second question: Can the inclusion of the field change the outcome of the minimax procedure, i.e., the value of the COV?

We consider the general case, depicted in Fig. 9. Piece 1 traverses a-squares in its trajectory; Piece 2, of the same or opposite color, sees Piece 1 within the horizon H_L. An investigation of the variation must show whether Piece 2 can arrive in sufficient time at its own trajectory's terminal square (a_f-square), which is an a-square of Piece 1's trajectory. This question arises if the two pieces are of opposite color; if they are of the same color, we should assume that Piece 2 can unconditionally arrive in time at

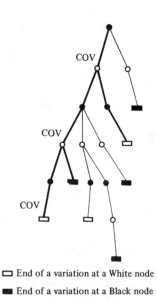

☐ End of a variation at a White node

■ End of a variation at a Black node

Figure 8 Which variations are to be improved?—marked by the thick line.

the *a*-square. Should we include in the play the field with the stem piece 2? The field may be one of control, blockade, or attack.

Clearly, the field should be included if the maximum profit from its inclusion can change the value of the COV. How is the answer to this question to be found?

We consider a control field with Piece 2 Black. Two cases arise: (1) there is an exchange on an *a*-square where Piece 1 stops; (2) Piece 1 goes through the square and participates in an exchange later on. In the first case, the Black Piece 2 is included in the exchange on the *a*-square, which may improve the value of the exchange for Black. If the profit from this improvement, added to the final value of the variation containing the exchange, exceeds the value of the COV, the field should be included in the mathematical model, since it may improve the value of the COV.

In the second case, the maximum expectation is that the White Piece 1 may be excluded from the exchange occurring further down the variation, after it has passed the *a*-square. This may lead to an improvement in the result of the exchange for Black. If this profit from the change in the result of the exchange, added to the final value of the variation, exceeds the value of the COV, it makes sense to include the Black piece in the MM. We may use the same method to decide the question of including a field of another type—a control field. We must remember that after the investigation of the subtree there may be such candidates for inclusion in the search, and in order to limit the size of the tree, we must establish priorities.

The determination of such fields, and the establishment of priorities for them, is one of the most delicate tasks to be accomplished by the algorithm.

Thus, given the conspectus resulting from the descent along a variation, we must, as a rule, build a subtree below the node where the question of inclusion of a field arises; otherwise the question cannot be resolved. In those cases, however, where there is no doubt that Piece 2 has time to arrive at the *a*-square and that the value of the COV can be improved (this can be determined by comparison of the value of the new target with the sum of the earlier derived values), the field must be included at once.

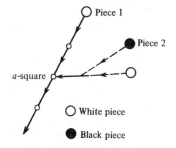

Figure 9   Whether to include a field of play.

The Repetition of Moves at a Node

We have already remarked that according to the canonical method for
forming a tree, there is no sense in repeating a move at a node: What sense
could there be in *re*-inspecting a node that has already been looked at? In
fact, however, repetition is senseless only when $H_L < 3$; then the trajectories
consist of a single displacement. Up to now, all known chess programs
embody such a horizon. If, however, $H_L \geq 3$ and trajectories may therefore
consist of more than one displacement, the repetition of moves at a node is
necessary.

Repetition takes on meaning since we attach significance to motion in a
trajectory rather than to a move. If the MM at a node is changed and a new
trajectory appears for which the first move coincides with a move already
contained in the MM and in which the piece has already moved, a repetition
of the move is not only sensible but unavoidable. The MM at the node
where the repetition occurs will be changed, but the current material value is
conserved.

The Depth of a Search and the Inclusion of Fields in It

Let us agree, to begin with, on a method for determining the limiting depth
D_L of a variation. As opposed to the horizon H_L, the depth may vary from
move to move. This is connected with the fact that a change in the horizon
has a very strong effect on the conservation of resources, whereas a change
in the depth has a delicate effect, since we have at all times the inequality
$D_L > H_L$.

We shall determine the depth by reference to the time that has been
expended in the prior portion of the game, i.e., by reference to the elapsed
time, and principally by reference to the time spent in contemplating the
preceding move. To a first approximation, we shall depend only on the
latter time. If more than three minutes were spent on the preceding move,
we decrease the value of D_L, and if less, we increase it. The extent of the
change will be determined by experiment.

The depth D_L of the search for the current move is therefore established.
We next decide how to use this value in selecting fields that are in the MM
for inclusion in the search. The unselected fields are to be frozen.

Let us assume that the search at the given node includes only those attack
fields whose targets can be annihilated (on stem trajectories) in the time
remaining before the limiting length is reached. If, for instance, this time
equals x plies and some two targets can be annihilated in a minimum total
time $t \leq x$, both the fields with these two targets may be included in the
search. If $t > x$ we can include only one of these two, or some other field for
which $t \leq x$.

This is logical enough; we are trying to win material within the truncated tree. Beyond the boundary, only positional estimates influence us. Therefore we include in the search only those fields in which we expect the target to be struck within the limits of the truncated tree.

Thus when the search reaches the limiting depth D_L, x naturally reduces to zero and the condition $t \leq x$ means that no further fields are to be analyzed in the search (strictly speaking, this condition may be encountered earlier; we shall return to this point later). If we are at the limit D_L, the search should continue further if there are fields of vulnerability, and in some circumstances (e.g., an attack on the King) also fields of partial vulnerability (when a retreat field is invulnerable or only partly vulnerable).

Let us now consider the case in which the condition $t \leq x$ is not fulfilled and the conditional depth limit has not yet been reached. Then, within the limits of the truncated tree, we cannot expect to win material via stem trajectories (if the field is not vulnerable, as we have just mentioned), and play on stem trajectories is therefore useless.

Then we should play on denial trajectories or on stem trajectories of a dependent field such that $t \leq x$ is satisfied. Although it would appear that this does not win material within the limits of the truncated tree, the appearance may be wrong if the play just described leads to making some stem trajectory vulnerable (with the reservation stated above). Then indeed the search may be continued beyond the depth limit.

If there are no vulnerable trajectories, and the condition $t \leq x$ is not satisfied for the remainder, the variation should be stopped at the depth limit.

In itself, D_L is not a depth limit for variations; it influences the depth only indirectly. It would appear that a chess master plays in this way—he has no explicit fixed length for a variation.

We are now in a position to answer some natural questions: What shall we do about breaking off a variation by reason of material? What targets should have their values included in the sum (Σm) of targets that enters the formula for breaking off? There should be no doubt about the answer: Include the values of only those targets that satisfy the condition $t \leq x$ (for stem trajectories and others in the field of play). These represent the greatest amount of material that can be won. We must unconditionally take into account the fact that the values of vulnerable and partially vulnerable targets are included in the sum Σm as a supplement. It follows that a variation is broken off on the basis of three factors: material, time, and the control of squares (vulnerability of the target). The formula for truncating branches (variations) is to be made precise in the same way.

We must now consider the difference between problems and practical games. In the search for a move from a position in a practical game, no limit is placed on the value of the optimal variation. Therefore the material relationship in the initial position does not influence the search. The

constructed problem is another matter; here we have a previously established task—win or lose. Therefore the initial material relationship must be
taken into account; it influences the values of the targets that must be
attacked and allows us to exclude certain goals from consideration, and
shortens the search.

The Pruning of Branches

In backtracking up a variation we are not absolutely required to develop a
search at a given node—to create a new branch at this node. We must
always answer the question: Is it possible to detach the as yet unconstructed
branch from those already formed here? The pruning of branches is in no
way to be confused with the stopping of variations. The stopping of a
variation (see Fig. 10) proceeds, in general, according to the formula
$-cm_T \geq \Sigma m_w + \Sigma m_b$, where Σm_w and Σm_b are the sums of values of
White's and Black's targets respectively; $c = +1$ at White's nodes; $c = -1$ at
Black's nodes; m_T is the current estimate made during the descent along the
variation; $m_T = M_w - M_b$, where M_w and M_b are the sums of values of
White and Black pieces, respectively, on the board in the current position.

A variation is broken off independently of the final value of other
variations, independently of the value of the COV (i.e., independently of the
results of the minimax procedure). The breakoff is organically related to the
goal of the play, to the value of the targets—and to nothing else.

The pruning of branches has a different meaning, although one case of
pruning has the same outward appearance as the breaking off of a variation.

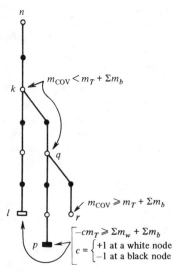

Figure 10   Whether to prune a branch.

Branches are pruned only to avoid useless work in the minimax procedure, i.e., branch-pruning is organically connected with the minimax procedure. The formula, at a White node, has the form

$$m_{COV} \geq m_T + \Sigma m_b,$$

and at a Black node,

$$-m_{COV} \geq m_T + \Sigma m_w,$$

where m_{COV} is the minimax value of the currently optimal variation. This formula applies the principle of expectation to the comparison between the value of the COV and the maximum value that may yet be obtained. If the new COV cannot surpass the old one, the branch must be pruned.

A branch at a node may be pruned at any point in its development. Therefore we may cut off the first branch at the node, thus cutting off the node itself (all its branches are gone), and outwardly this appears to be no different from stopping the variation. Pruning may be initiated at any branch.

On the left-hand side of our formula for pruning stands the value of the COV. This is the largest value obtainable in the variation and, in particular, may be obtained at the node where the pruning is under way or at a node higher up in the variation.

Let us illustrate this by an example displayed in Fig. 10. We descend along the variation from the node n to the node l, where the variation is broken off, for the same reason as at node p. In backtracking to the node k, it becomes clear from the mathematical model that a second branch can be constructed there. We make use of the pruning formula; the sign of the inequality is reversed, and the branch must be constructed. We descend to the node p, where the variation is ended; we backtrack to the node q, where the pruning formula gives a negative result. Then we descend to the node r, where the pruning formula tells us to stop constructing the branch.

There remains only to define the quantities Σm on the right-hand side of the pruning formula. This is not a simple matter. It would appear that they must contain the values of the targets (both included and not included, since the latter may appear further down the branch) and the strikable fields. The final answer, however, can be obtained only by experiment. Among other things, the same could be said of the formula for stopping a variation. On the right-hand side, the sum $\Sigma mw + \Sigma mb$ should not contain all Black and White target values, but which ones it should contain must depend on experiment.

Three States of a Field (The Active Mathematical Model)

The MM changes during the search. New fields of play arise, the MM expands, the amount of information to be processed increases, and this brings about a substantial slowdown in the operation of the program. Therefore a field should be included in the MM only when necessary, and should be excluded when no longer needed. From this point of view, a field may encounter in the course of its life the following states:

1. It is included in the MM. It takes part in all the affairs of the MM: the positional estimate, the determination of priorities, forking, etc. Information on the fields included in the MM is gathered and saved.
2. The field is excluded from the MM during the descent along the variation. This may occur as a result of the following causes: an a_0-piece is annihilated or leaves the a_0-square without occupying the a_1-square; or the same thing happens at the a_f-square when we are not dealing with a stem attack trajectory. Under these circumstances we have said that the field should be frozen and all information about it removed from the data immediately connected with the MM.

    We must remember that the frozen data, which are not needed for the current processing operations, should be unfrozen again during the backtracking (at the node immediately above the one at which they were first frozen) and included in the MM at the necessary moment.
3. During a further backtracking, when we have passed the node at which the field was first found during the descent, the field should be erased from the memory of the computer; in the general case it will not be used again.

Thus the MM contains only those fields described in case 1 (see Fig. 11). This leads to a sharp cutback in the MM; it can vary from node to node, during both descent and backtrack. Such an MM we shall call "active". We should note that everything we have said about frozen fields also applies to frozen denial trajectories—they are to be stored in the same place as the similar fields. With respect to trajectories, however, the matter is somewhat more complicated: already frozen denial trajectories of fields that are still included in the MM must be inspected whenever we solve the problem of

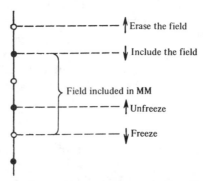

Figure 11   Field states. Backtracking: up the variation; Descending: down the variation.

forming a new denial trajectory, so that we may avoid the construction of an existing frozen trajectory. Thus the fields of both types 1 and 2 are subject to recall; but information processing pertains only to fields of type 1.

The active MM changes from node to node during the search. The MM is used to determine the positional estimate and the priorities for inclusion of trajectories in the search. It plays an exceptionally important role at the nodes where the variation suffers a compulsory cutoff; we shall return to this point.

Keeping the MM in Mind at a Terminal Node

When a move from the initial position is made in accordance with the optimal variation established by the minimax procedure, we should not recommence the search from the new initial position if the opponent's answer was in accord with his move in the optimal variation. If we were to renew the search at the vertex saved in the subtree, we would be wasting effort. (Note that the unselected portion of the subtree, which was unnecessary, has already been erased from memory.)

Let us first of all convince ourselves that the preceding optimal variation (or variations) will remain optimal in the new position. We expect it to do so, since the opponent is playing according to this variation. We go down to the terminal node of the variation and recommence the search in the subtree at that point, which has been saved in memory. To launch ourselves into the search, we need the mathematical model which was developed there, and therefore it should be available in memory. So, while forming the search tree, we must save those MM's that correspond to terminal nodes of the optimal variation or variations.

A Technical Question

Let us consider one of the technical questions that may also influence the construction of a small tree: if in a variation there arises a position that has already been assessed in another variation, it may be that further work on the new variation is senseless, since we already know the value of the position itself. We must, however, verify that not only the positions but also the mathematical models coincide. We can do this indirectly by testing the coincidence of the priorities for including moves in the search.

When a position is repeated in the same variation in which it arose, the rules of chess imply that the variation should not be continued—it is meaningless. We may define the repetition of positions by means of a library of current positions; this library will be small, since the search tree is.

B. The Use of Historical Experience

When a chess master plays a game he uses historical experience (his own and the experience of others) in four different ways:

1. *In parrot fashion*. This is characteristic of the opening. The theory of the opening is not subject to judgment in any of its parts, and the master makes his moves in the opening without inserting himself into the process, i.e., he moves as a parrot talks.

2. *By the handbook method*. In playing a game, the master seeks out, in his store of accumulated knowledge, the exact position occurring in the search. This library position has a score; since the positions coincide, the score of the variation is known, and the variation itself may be cut off. This usage of historical experience is characteristic of the endgame, but may be successful in the opening as well, when moves are transposed.

3. *By the outreach method*. This is based on an attempt to reach positions that the handbook method can use. The master looks up library positions having a favorable outcome and lying near the search position. (See the Glossary of Terms for a definition of nearness.) Having found such positions, the master constructs his search so as to get them into the search tree if possible. Then he relies on method 2. Therefore this method is also characteristic of the endgame and, perhaps, of the opening.

4. *By the associative method*. This is based on a partial congruence of the position in the search with positions in the library. The master seeks out a fragment of a library position, i.e., a small group of pieces whose action has led to success. If the search position contains the same group of pieces in the same constellation, he includes the group in his search, in order to see whether it will lead to success in the current case. If the fragment has often been successful in the past, it is apt to succeed again. This method determines a direction for the search and on the average it saves resources while the search is under construction. It would appear that the associative method presents the only way to apply historical experience to the middle game and to complex endgames.

The Library of Openings

In PIONEER (our chess-playing computer program based on the algorithm we have presented) the library of openings is small and the program uses the parrot method, which is employed in all chess programs that have a library of openings, and bears only weakly on the essence of the problem.

The variations in the opening are limited to 12 plies. It is assumed that the search tree is constructed from that depth onward. The list of identified openings is also kept to a minimum.

An expansion of the library of openings is being contemplated; in principle, PIONEER could learn to fill out its library on the basis of experience gained in the games it plays, and possibly from the games of others.

The Library of Middle Games

There is no doubt that the associative method is the most complicated way of using knowledge accumulated from history. We have to enter fragments in the library that will direct our search. To this end we divide the pieces in a fragment into two groups: a group of fixed pieces, standing on predetermined squares, and a group of pieces attached to predetermined squares but standing at some distance away. The fixed pieces are not intended for inclusion in the search; the side using the fragment is not interested in them. The attached (the so-called bound) pieces should be trying to reach the squares they are attached to.

One fragment may define a set of fragments, if it may within reasonable limits be moved around on the board. This should be kept in mind when forming the library. If a group of pieces in a position occurring in the search tree is congruent to the position of the same pieces in a fragment, that fragment determines the priority for the inclusion of the attached pieces in the search—that is, it directs the search.

For the employment of the associative method, a complex endgame is not in any way different from the middle game. Consequently, the endgame library should be constructed in fragment form. Up to now, this has not been done.

The middle-game library is not large; it contains about 70 fragments, which by displacement around the board generate some 630 positions. In principle, it can be self-expanding, using the principle of self-learning.

The Endgame Library

From the chessplayer's viewpoint, filling the endgame library is a simple matter, since endgame theory, as opposed to middle-game theory, deals with a large set of positions. From the programmer's viewpoint, the task is complicated.

To simplify the task, we assumed that the use of the endgame library would involve only the handbook and outreach methods. Strictly speaking, the handbook method may be regarded as a particular and very simple case stemming from the method of directed search.

Endgame theory, as usually presented by those who write about chess, is connected with the choice of variations. This view can hardly be used for programming, since chess masters do not use it. We assumed that the library contains positions marked with the side that has the move, the value (win, lose, or draw), and when necessary, the recommended move.

For compactness of the entries in the library, the positions were described by templates in which the White King was set on some certain square of the board. The program includes formulae for computing the value and the recommended move when the coordinates of the White King are changed within admissible limits. In this process the relative coordinates of the remaining pieces are not changed. We could have stored all the necessary positions simultaneously in the computer memory; then the displacement formulae would not have been needed, but a large amount of memory would have been occupied.

Some 700 templates were stored in memory; these generated about 7000 positions. Taking account of the assignment of the move to one side or the other, the total is 14 000; the author knows no more than half of these positions. Moreover, taking account of vertical and horizontal symmetry, and (in the absence of Pawns) diagonal symmetry, we obtain some 35 000 positions.

These belong to the so-called technical endgame, a knowledge of which constitutes an integral part of the master's chess technique. The number of pieces (8) was provisionally limited, and the positions were divided into classes by reference to their material (in all, 31 classes). After the library was established it was tested, by parts, by ten expert players. A defect was observed in about 10% of the positions. After our corrections, we may assume that the number of defective positions scarcely exceeds 1%, which is acceptable.

Some restrictions were also imposed on the search for a position: (1) pieces in the search position must agree identically with the pieces in the library position; (2) the coordinates may disagree for at most two pieces; and (3) the total number of moves required for these two pieces to coincide in position must be less than a specified limit.

The side that is trying to find a position forms a sheaf of stem trajectories (germs of fields) and if necessary a field. The trajectories of other pieces are not formed; they may exist only if they are already formed for other reasons. If they are so formed, then in principle the attempt to reach the library position is admissible. For the enemy pieces, these trajectories will be anti-forked in the sense that the enemy must act along them only in extreme circumstances.

The endgame library should contain not only positions, but also rules, e.g., the rule of the square. However, only by experiment can we determine what rules should be included.

Stocking the endgame library by a self-learning process is an increasingly urgent long-term task. If results are achieved in this direction, they may well apply to other and more delicate aspects of self-learning. Work has already begun on self-stocking of the library.

In a chess game, a master continually refers to historical experience in the opening, the middle game, and the endgame. Although he does not apparently do this at all nodes of the tree, it nevertheless follows that the tree cannot contain many nodes, and the program for using the libraries must be so organized that it requires only a small expenditure of computer resources.

Let us consider the problems of priority connected with the middle-game and endgame libraries. If we neglect the case in which the search and library positions coincide completely, so that the variation is broken off, we see that the presence of the library clearly may lead to a change in the adopted priority order and to the appearance of new trajectories in the MM.

In fact, our priorities were adopted on the basis of the average experience of the past; when they are applied, the search tree will, on the average, be at its smallest. There are, however, exceptions to the rule: if in the past we have encountered positions where the search tree could be sharply reduced by using the rule of priorities, such a position (fragment) should be included in the library, and this notion should be the foundation of the self-stocking and direct-stocking processes. The use of these library positions implies a reordering of the priorities like that which took place in the past; all the priorities unaffected by this rearrangement are conserved. Each position is to be assessed by reference to the value of the targets, so that we may use this information in determining the priorities.

In comparing target values (including the values assigned in the library) we must keep in mind another provisional value, which we shall take as decisive. For instance, we may suppose that in an equilibrium position, differences in material amounting to two or three units will decide the outcome of the game. Therefore it we can play on a vulnerable attack trajectory against a target of less value than the one contained in the library, but such that we can gain a decisive value, the priority for including the library position in the search is postponed.

CHAPTER 5
Three Endgame Studies
(An Experiment)

Even before the work on the program began, it was decided that the first trials would be made on chess problems. The solution of a problem by the program would be a very profitable experiment, and at the same time not too complex. One of the bases for a chess master's strength is the calculation of variations; after this master capability had been formalized in the form of a program, it was subjected to an experimental trial by endgame studies. Some eleven compositions were prepared, but it turned out that we secured the information needed after only three experiments.

Figure 12   A problem due to Richard Reti. Draw.

Figure 13   Problem due to Botvinnik and Kaminer. White to win.

It was assumed that the studies would represent an easy output for PIONEER, since they contained no positional niceties and all the variations could be pursued to the end. It was also assumed that no library of openings, middle games, nor endgames would be needed. In reality, things turned out to be much more complex.

On the first problem, a study by R. Reti (Fig. 12), PIONEER stumbled. The search tree grew, the variations did not reach an end, and PIONEER did not know which moves had priority. A chess master, however, knows when to break off a variation and which move to prefer for inclusion in the search—then his search tree is small indeed.

It was necessary to put rules into the program; these were founded on the well-known rule of the square. Then the search tree reduced to 54 nodes. If we had possessed a library of current positions, we could have avoided the repetition of variations, and the number of nodes could have been reduced to 45 (Fig. 14).

Some variations appear to be unfinished, but in fact they were carried to their logical ending—they were scored by modifications of the rule of the square that were put into the program.

PIONEER found difficulty in a problem by Botvinnik and Kaminer* (Fig. 13), despite the fact that it is very simple (I was 13 at the time we composed it and Kaminer was 14). It again turned out that if moves in the trajectories of the mathematical model are included in the search without analysis, PIONEER's search tree differs markedly from the one a master

*This problem has an amusing history. In 1925, when Sereža Kaminer and I developed it, I proposed that there should be a Pawn on square g6, but Sereža insisted that it should be a Bishop. He persuaded me, and the study was published with his text. In January 1977, in recalling this composition, I wrongly placed a Pawn on g6, and PIONEER solved the problem on the first variation.

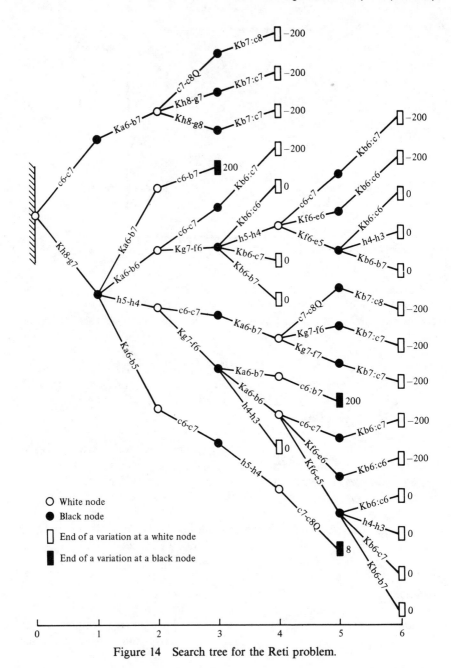

Figure 14   Search tree for the Reti problem.

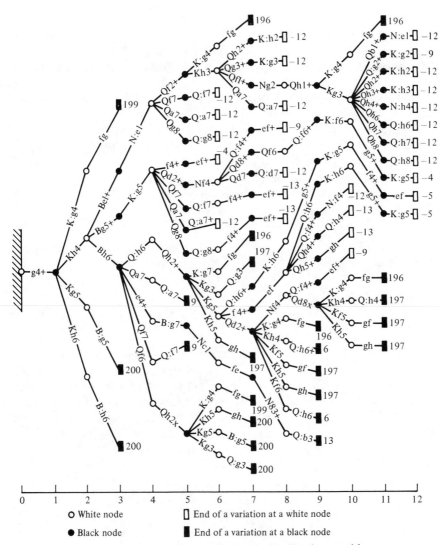

Figure 15   Search tree for the Botvinnik–Kaminer problem.

would produce. It was necessary to prohibit play in fields where the value of the targets was less than the loss of material and to introduce rules for terminating a variation. Apparently these difficulties arose from the lack of a positional estimate. The search tree, in the end, consisted of 145 nodes (Fig. 15).

It appeared that we should postpone further attempts at a solution until a faculty for positional estimates was introduced. This we did not do, however. The fact is that the Reti problem contains so few pieces, and these are so nearly immobile, that it is completely soluble by programs using a full-width search. Although the Botvinnik–Kaminer problem contains more pieces, including some long-range ones, there are so many captures and checks that it too is soluble by programs using full-width searches, since these carry a variation beyond its standard length when there is a capture or check. We therefore attempted to solve a problem due to G. Nadareišvili, since there was no doubt about its complexity, nor about the inability of other programs to overcome it (Fig. 16).

The basic variation discovered by PIONEER is:

1. g6 Kf6; 2. g7 Bh7; 3. e4!! Nf3; 4. e5 + N:e5; 5. K:h7 Nf7; 6. g8Q Ng5 + ; 7. Q:g5 + K:g5; 8. h6 c4; 9. Kg7 c3; 10. h7 c2; 11. h8Q c1Q; 12. Qh6 + Kf5; 13. Q:c1.

At the outset the program took so much machine time that it practically precluded an arrival at a solution. Then B. Stilman was compelled to act like an aeronaut whose balloon is losing height at a menacing rate—he must throw out not only ballast but alas! even useful cargo. Thus, Stilman threw out the procedures for the deblockade of trajectories (leaving only the deblockade of pawns), and for this reason the reader will not find one of the author's variations. But we did succeed in speeding up the solution to such

Figure 16   Problem due to G. Nadareišvili. White to win.

an extent that less than four hours were required. To play at master strength, PIONEER needed a faster machine.

The program was not completely clean; there were more than a few technical errors that could have been found and eliminated by further experimentation, but that too would have needed a faster machine. For this reason, a very strange move (Nd7-g7) wormed its way into the search tree, and there were other mistakes. But even this draft version of the search tree (Fig. 17: pp. 60–61) means a great deal, since there are, after all, 200 nodes in the tree.

It is true that during the tests we patched in some palliative rules for positional estimates, but we did draw some extremely important conclusions. It was only after this work that we succeeded in properly creating the priorities for the inclusion of fields in the search.

The logic of Nadareišvili's problem consists basically in the fact that Black's King, on square f6, controls the retreat square of White's King; this and only this gives Black a counterplay. It is hardly possible to construct a humanoid tree without understanding this. The niceties of White's play are developed in part to drive Black's King off square f6. Our earlier analysis of part of the subtree, with the corresponding variations, may be useful here. Comparing the results of an analysis of two portions of the subtree, we perceive the role played in the COV by Black's King on f6. Formally, we would call this "understanding".

We should look at the craftiness of Nadareišvili's problem from the viewpoint of a chess practitioner. A master always looks for vulnerable and strikable targets; on the third move of the COV the Bishop at h7 is such a target. After he has been annihilated on h7 a vulnerable target is the Queening of the Pawn at g7. The attack by the Pawn at e3 on the King at f6 is controlled by Black on e5, and Black's King can retreat. All this is a psychological barrier that must be broken in order to arrive at a correct solution. Play in fields other than those with vulnerable and strikable targets is required!

Now that the program is basically completed, further experiments are required for its perfection.

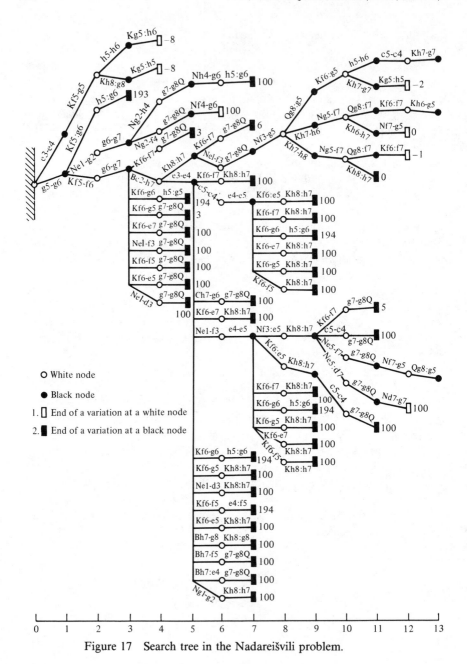

Figure 17   Search tree in the Nadareišvili problem.

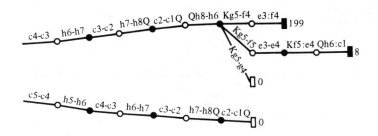

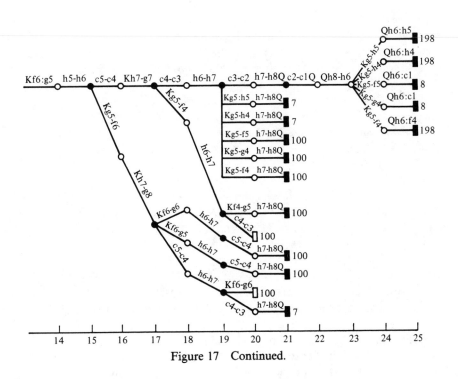

Figure 17   Continued.

CHAPTER 6
The Second World Championship

The First World Championship contest among chess playing programs took place in Stockholm in 1974, under the aegis of IFIP (the International Federation for Information Processing). The winner was the Soviet program KAISSA.

The history of chess tournaments among computers begins in 1970, when the first U.S. Championship was held in New York, timed to coincide with the annual meeting of the ACM (Association for Computing Machinery). Since then, these championships have been conducted annually.

World championships for computers are held triennially, as they are for humans. The second was held in August 1977 in Toronto, Canada.

These contests demand a substantial financial outlay, in machine time and communication channels. Therefore they are usually conducted on the Swiss system, in four rounds. The time limit is set at 20 moves per hour.

KAISSA (whose authors are G. M. Adelson-Velsky, V. L. Arlazarov, and M. V. Donskoi) used a full-width search in both tournaments. Its chief opponent—the program CHESS 4.6 of the USA—tried in 1974 to avoid a full-width search and was unsuccessful. In 1977, the Americans (David Slate and Larry Atkin) returned to the full-width search and won the championship.

Both these programs stand out among others; in all there were 16 contestants. They differ in detail, but are basically similar. Both take account of all variations up to the limiting length, using the so-called branch-and-bound, or α-β-cutoff, method; the variations are carried further if there are captures or checks.

A demonstration game, played after the championship, gives a clear picture of the games characterizing the strongest programs.

KAISSA–CHESS 4.6

1. e2-e4 Nb8-c6

This little-known opening, due to Aron Nimzovitch, was doubtless planned by Black in order to avoid troubles that might come from KAISSA's library of openings.

2. Ng1-f3...

White was obviously not prepared for this opening; the theoretical move is 2. d4.

2. ...e7-e6

Black continues to avoid known continuations which would arise after 2. ...e5.

3. d2-d4 d7-d5
4. Bf1-d3...

A move that would hardly be made by a qualified chess player. With ...Nb4, Black could exchange the Bishop at d3, after which his opening difficulties are behind him. After 4. e5, Black has difficulty.

4. ...d5:e4
5. Bd3:e4 Bc8-d7
6. 0—0 · · ·

A stereotyped development: 6. Ne5 N:e5; 7. de suggests itself, and White's positional superiority is evident.

6. ...Ng8-f6
7. Rf1-e1...

White conserves a freer game, but in parting with the King's Bishop, he reduces his active opportunities. 7. Bd3 suggests itself.

7. ...Nf6:e4
8. Re1:e4 Bf8-e7
9. c2-c4 f7-f5

A scantily motivated weakening of the squares in the e-file and in the diagonal a2-g8.

10. Re4-e1 0-0
11. Nb1-c3 f5-f4

This move has an extremely unfavorable effect. Even the simple moves 12. Ne2 g5; 13. d5 leave Black in a critical condition because of the weak position of his castling; 12. Ne4 would also have been sufficient.

12. Qd1-d3 Qd8-e8
13. g2-g3...

Although KAISSA, aside from a few small adventures in the opening, had stood up with dignity, this move dispels all illusions. Black's initiative in the f-file assumes a menacing character; the weakness of the square f3 is especially painful. After 13. d5, Black would have a difficult problem because of the poor placement of his Queen on e8.

13. ...f4:g3
14. h2:g3 Qe8-f7
15. Bc1-f4 g7-g5

This looks very dangerous, since after 16. N:g5 B:g5; 17. B:g5 Q:f2 + ; 18. Kh1 N:d4; 19. ...Bc6 is inevitable.

16. d4-d5... (see Fig. 18).
16. ...e6:d5

After 16. ...Nb4; 17. Qe4 gf; 18. de B:e6; 19. Q:e6 fg (or 19. ...Nc2; 20. Q:e7 Q:e7; 21. R:e7 N:a1), Black would have a material advantage.

17. Nc3:d5 g5:f4
18. Nd5:e7 + Nc6:e7
19. Qd3:d7 Ne7-g6
20. Qd7:f7 + Rf8:f7

The complications have disappeared and a tranquil endgame has arrived. Up to this point the combat was equal, which should be especially noted: KAISSA used a computer with a speed of 3×10^6 operations per second; CHESS 4.6 had a speed of 12×10^6 operations per second. KAISSA could calculate variations to a depth of five plies; CHESS 4.6 could go to six. This

Figure 18   Position from the demonstration game KAISSA–CHESS 4.6.

is a small difference, which turned out to be insignificant. But, as pieces are exchanged, the number of possible moves and positions decreases and therefore so does the number of nodes in the search tree; the length of a variation can be increased. KAISSA could handle 90 000 nodes in the tree; CHESS 4.6, 400 000. In the end, KAISSA extended the length of a variation to nine plies and CHESS 4.6 went to twelve. In the endgame, CHESS 4.6 was the stronger. KAISSA's tactical errors consisted in not avoiding exchanges.

21. g3-g4...

A good move. As the authors of KAISSA remarked, the program was trying to maintain Black's isolated pawns in the f- and h-files.

21. ...Rf7-d7
22. Ra1-d1 Ra8-d8
23. Rd1:d7...

Useless. After 23. Re8 + R:e8; 24. R:d7 Re7; 25. R:e7 N:e7; 26. Kf1 with the sequel Kf1-e2-d3-e4, White would have a clear advantage in a Knight ending because of the weakness of the Pawn at f4.

23. ...Rd8:d7
24. Kg1-g2 Kg8-g7
25. Nf3-g5...

Beneath criticism. One must not let Black's Rook into the second row.

25. ...Rd7-d2
26. Rc1-b1...

Possibly bad also would have been 26. Ne6 + Kf6; 27. N:c7 R:b2, and the White pieces are isolated.

26. ...Rd2-c2
27. b2-b3

Doubtful also are 27. Ne6 + Kf6; 28. N:c7 R:c4.
27. ...Ng6-e5
28. Rb1-h1 Rc2:a2
29. Rh1-h4....

The endgame is hopeless, and after 29. R:h7 + Kg6; 30. Re7 Nc6, White loses a piece.

29. ...Ne5-d3
30. Ng5-h3 Ra2-b2
31. g4-g5 Kg7-g8

A simpler move would have been 31. ...a5.

32. Nh3:f4...

An erroneous estimate of the Pawn ending. The last chance was in the continuation 32. Kf3 R:b3; 33. N:f4 Ne5+; 34. Ke4 N:c4; 35. Nd5. The rest is obvious:

32. ...Rb2:f2+; 33. Kg2-g3 Rf2:f4; 34. Rh4:f4 Nd3:f4; 35. Kg3:f4 Kg8-f7; 36. b3-b4 Kf7-e6; 37. Kf4-e4 a7-a6; 38. Ke4-f4 Ke6-d6; 39. Kf4-e4 c7-c5; 40. b4:c5+ Kd6:c5; 41. Ke4-d3 a6-a5; 42. Kd3-c3 a5-a4; 43. Kc3-d3 Kc5-b4; 44. Kd3-c2 Kd4:c4; and White resigned.

Among the other competitors, Monroe Newborn's OSTRICH deserves special attention. It was the only entrant to act on its own, without communication lines; it operates on a microcomputer, which sat on a table beside OSTRICH's author during the games. By an irony of Fate, the computer went out of order during the third round; with a winning position, OSTRICH was counted as defeated. (In the fourth round, a replacement for the computer was found in Montreal.) Although OSTRICH uses a small machine, it plays on a level with other programs. Newborn plans to use it as a basis for defining concepts that will lie at the foundation of a spectrum of programs for micro- and mini-computers.

A championship contest among computers is an engrossing spectacle. The authors of the program sit at chess boards. Using video terminals, they inform their own distant computers of the moves made by the opponents. The computer's answer is displayed on the terminal and echoed on the chess board. While waiting for the answer the scientists visit amicably among themselves, analyze positions, argue, joke, and often criticize the performance of their own programs. This is all quite understandable: a computer tournament is only formally a sporting event—in essence, it pursues scientific aims.

After the tournament had ended, and with it the decisive match KAISSA-CHESS 4.6, David Cahlander (a consultant to the Control Data Corporation, whose CDC computer was used throughout by the American program) telephoned Minneapolis and fed Nadareišvili's problem to the Cyber-176. CHESS 4.6 found the first two moves correctly, but as soon as it made the third move, Cahlander glanced at the display, burst out laughing, and threw up his hands: the program had examined about a million moves and not found the right course.

After the match, Slate and Atkin, the authors of CHESS 4.6, remarked that they intended to abandon the brute force method (the term used in the USA for a full-width search) as having no future, and turn to an evolutionary method of program development. How much time will it take to re-traverse the path followed by PIONEER?

In the brute force method a higher computer speed yields only an insignificant increase in the depth of the calculations, and only a modest increase in chess-playing strength; in PIONEER, a higher speed yields a proportionately greater depth of the calculations. In the meantime, PIONEER plays slowly. The solution of Nadareišvili's problem would take less than ten minutes on a machine of the Cyber-176 class. By no means could all chess players, even of the highest qualifications, manage it in that time.

The third World Computer Championship will be held in 1980. A qualifying tournament will take place in Japan, after which the two winning programs will compete with CHESS 4.6 and KAISSA in Australia, where the third World Champion will be selected. We may expect that this champion will even now play like a chess master.

At the end of the Toronto championship, the competitors—the authors of the programs—took part in a conference at which the Dutch programmer B. Swets called for the formation of an ICCA (International Computer Chess Association). This was approved in principle. A sign of the times!

Conclusions

Since 1964, when the author began to seek support for his algorithm, there have been many critical remarks; these are worth reviewing. It has been said that this is all fantasy: it conflicts with the accepted canons; it demands more resources than does a full-width search; it would require decades for development; a collective employing no less than 20 mathematicians would be required for the programming; the capabilities of current computers are clearly inadequate for the implementation of such an algorithm; and so on.

However, 14 years have passed (and work on the program began only in 1972), and we are on the threshold of solving a great scientific problem. An algorithm that models the behavior of a chess master has been implemented (and the generally accepted methods are looking more and more doubtful); the work has gone forward with a minimal number of programmers; the capabilities of today's powerful computers have turned out to be wholly adequate.

Archimedes, who founded the theory of the lever, said, "Give me a place to stand and I will move the Earth." Perhaps the reader will agree that it is now appropriate to paraphrase his dictum as, "Give PIONEER a fast computer and the theory for solving inexact problems will assist in better management."

APPENDIX 1
Fields of Play

B. M. STILMAN

The highest level of the control system in Botvinnik's algorithmic model of the game of chess is the mathematical model (MM), which is an ensemble of fields. In the process of searching for a move, the MM continually changes itself by including and excluding fields, and thus directing the search. In this appendix we consider the problems of (a) forming the included fields and (b) searching for moves in the ensemble of fields that are included. The MM is kept invariant with respect to the composition of the fields it includes; its formation and constant reconstruction are not considered here.

All the applicable algorithms are described concisely. Special attention is given to their implementation in PIONEER. It is worth noting that many of the procedures described below were worked out in detail and refined during the development of PIONEER and in the experiments conducted with it.

1. The Formation of Fields and the Search Within Them

Pieces in motion along a trajectory try to occupy its terminal square (a_f-square). Suppose that in order to do this, an enemy piece standing on that square must be annihilated. A second-level subsystem must be organized so that enemy pieces oppose the motion of the attacking piece in its trajectory and friendly pieces support it. Since all these move in their own trajectories, an agreement can be reached that will put the play along the several trajectories under the control of the overall goal of the ensemble of trajectories. The ensemble of trajectories and pieces entering the local combat, divided into two hostile camps, makes up a field of play.

1.1. The Concept of a Field

Suppose that the a_0- and a_f-squares contain pieces of opposite color $(+)$ and $(-)$, called an a_0-piece and a_f-piece, respectively. The assertion that the a_0-piece attacks the a_f-piece means that there exists a trajectory leading from the a_0-square to the a_f-square and that the time (in half-moves) required for the a_0-piece to complete its trajectory does not exceed some number H_L, which is called the *limiting attack horizon*. The contemplated trajectory of the $(+)$ piece is called a *stem trajectory* of the field, and the a_0-piece a *stem piece*.

Let us consider the set of squares in the stem trajectory: they consist of a-squares, where the piece comes to rest, and b-squares traversed without a halt. The set of trajectories of the $(-)$ and $(+)$ pieces terminating in these squares will be called *first-order denial trajectories*. The set of trajectories ending on squares in first-order denial trajectories will be called *second-order denial trajectories*, and so on.

Such a set of Black and White pieces, and the set of trajectories defined in the manner just described, will be called a *field*. Its color is the color of its stem piece. The precise meaning of this notion will emerge when we describe the process for calculating the trajectories to be included in the field.

We shall show how the fields of play are formed, using a specially developed artificial position (Fig. 19).

First, however, we must see how PIONEER searches for a move.

Figure 19   An artificial position.

1.2. The Search Procedure in the Initial Position

The search tree is depicted with its root at its top, i.e., the variations in the search leading from the initial position grow their branches downward. We search down along the tree from the initial position by a sequence of moves. By some criterion we end the variation, and the final position is scored by some chosen scoring function. Then we backtrack by one move, ascending the tree and, remembering the position below, we score the position just reached. From here we descend again by moving along another branch. The variation is again ended, and again we backtrack, lifting the score with us, make another move down the tree in the variation, and so on. After we have explored all possible ultimate moves in the branches of the model, we repeat the process with the penultimate move, and so on. The scores derived from below are compared; for Black's moves the minimum score is selected, and for White's moves the maximum.

A flowchart of the search for a move in the initial position is depicted in Fig. 20.

Downward motion in the search tree is controlled by the procedures shown on the right hand side of Figure 20. The upward motion, necessary for the minimax procedure and for continuation of the branches, is controlled by the procedures shown on the left hand side of the flowchart. The symbol D denotes the current depth of the search tree, in plies.

In the example considered below, for demonstration purposes, only the shaded blocks of the flowchart are activated. They control the formation of a field during the search process, and are described in detail later. The remaining procedures construct the mathematical model.

1.3. The Formation of Fields During the Search—An Example

Consider the example shown in Fig. 19. In this position we set up a program with a close horizon, equal to two moves; therefore it can construct only one White attack field, that of the Bishop at f2 against the Pawn at e5. Any other possible field lies outside the horizon, and the program does not construct it. Thus, in the initial position, we compute the Bishop's trajectories f2-g3-e5 and f2-d4-e5, which make up the so-called *stem sheaf*. Then the program begins the search for a move (Fig. 21). A precise description of all the procedures illustrated in this example will be found below.

Since Black is to move and we have no Black trajectories, i.e., no moves for him, we allow him to make a null move. Then it is White's turn, and the program makes the move 2. Bf2-g3 in the stem trajectory. Then it begins a search for possible opposing trajectories on Black's part and support trajectories for White, i.e., control trajectories focussed on g3. It looks for Black trajectories that will allow a piece to arrive in time to join the battle on g3; in this case, the length can be two displacements. In the White field

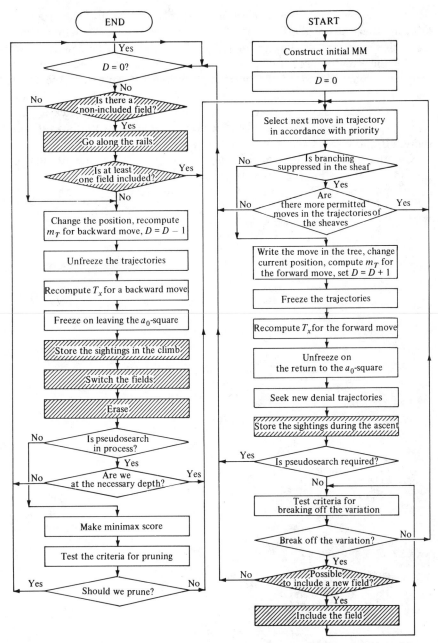

Figure 20   Flow chart of the routine for seeking a move in the original position.

the support trajectories must consist of one displacement only. The program finds the sheaf of trajectories g7-f5-g3 and g7-h5-g3 for Black's Knight, and for White's support, the Pawn move h2-g3. Having found the new trajectories for the Knight, along which motion would be senseless since he would arrive too late, pursuit of the current variation is terminated. We must backtrack along the variation by one half-move in order to include the Knight's trajectories in the play. In our case this means going back up the tree to the initial position. Therefore the program ends the variation without scoring it and undoes all its moves in reverse order.

Note that in the printout of the search tree (Fig. 21), the indentation of the move in the listing is proportional to the depth of the move in the tree, i.e., the number of half-moves in the variation leading from the initial

```
WHITE *KB1,BF2,PG4,PH2,
BLACK *KA8,NG7,PE5,
 BLACK TO PLAY


******
      BF2-G3
      BF2-G3   500
****** 500
NG7-F5
      PG4*F5
      PG4*F5    3
NG7-F5    3
NG7-H5
      ⁵G4*H5
      PG4*H5    3
NG7-H5    3
******
      BF2-G3
            ******
                  BG3*E5
                  BG3*E5   500
            ****** 500
      BF2-G3   500
****** 500
NG7-H3
      PG4*H5
      PG4*H5    3
NG7-H5    3
NG7-F5
      PG4*F5
      PG4*F5    3
NG7-F5    3
NG7-E6
      BF2-G3
            NE6-F4
                  BG3*F4
                        PE5*F4
                        PE5*F4    0
                  BG3*F4    0
            NE6-F4    0
      BF2-G3    0
      BF2-D4
            PE5*D4
            PE5*D4   -3
      BF2-D4   -3
NG7-E6    0

      SIZE OF THE TREE   13
      SIZE OF THE MAP    26
      TOTAL NUMBER OF MOVES CONSIDERED IS 18
      CPU TIME IS    25 SECONDS
```

Figure 21 Printout of PIONEER's search for moves in forming fields. (Standard algebraic notation; * * * * * * means the move is skipped.)

position to the move itself. The number 500 to the right of a move signifies that the variation has ended without a score.

The program has returned to the initial position. In the return along the unscored variation, the whole tree as constructed was erased from memory. The program begins to descend a new tree, but now has Black moves in trajectories. It registers the move Ng7-f5 in a control trajectory. Here, however, the program finds the Pawn trajectory g4-f5. Since at this point in the play this is a sensible move, the program does not back up along the variation. First of all, in this variation it makes an extremely profitable capture, and therefore makes the move 2. g4:f5. Here the variation ends, since Black has lost three units of material (the value of the Knight), which is more than he expected to save (one unit, the value of the Pawn at e5). The difference between the White and Black material removed from the board in this variation is equal to three, which is the score for the variation. The program now backtracks up the variation with the score 3 (written to the right of the description of the move in Fig. 21).

In the position arising after the backward move corresponding to the capture 2. g4:f5, White has an as yet unstudied move (the Bishop's move). The program ascertains that the lifted score 3 is as bad for Black as the preceding move (1. ...Ng7-f5); they are equivalent in value as seen by the minimax procedure. Therefore, in the given position, White sees no point in pursuing other moves and the program backtracks up the variation. As a result, we arrive at the initial position. We begin a new variation with the first move 1. ...Ng7-h5 on the trajectory g7-h5-g3, and are again returned to the initial position. Thus in the initial position all Black moves opposing the attack of the Bishop on the Pawn result in a score favoring White, i.e., for the present no satisfactory defence can be expected. The search continues: Black omits a move, White answers with 2. Bf2-g3. No new trajectories have been found, but the old Knight trajectory from g7 to g3 remains. The program finds that the time available for the Knight's movement in its trajectory is exhausted, so that the trajectory makes no sense. The Knight cannot take part in the control of the square g3.

Thus Black has no admissible moves; he omits a move, and White makes the move 3. Bg3:e5. In this position the program finds a blockade trajectory for the Knight from g7 to f4 (g7-h5-f4 and g7-e6-f4). The square f4 is a so-called b-square of the Bishop's trajectory f2-g3-e5, i.e., a square traversed by the Bishop without halting. The search for a Black control trajectory focussed on e5 was unsuccessful; the program found no such trajectory of length not exceeding three half-moves. (We note that in a position differing from that shown in Fig. 19 by the position of the Black King, which stood on b7, the program found trajectories for it from b7 to e5 and subsequently included them in the play.)

Since a new trajectory, for blockade of a b-square, has been found, we must backtrack along the trajectory by as many half-moves as make sense in the motion along it. The program ascertains that in the present case we must return to the initial position. The return is made without a score: all scoreless backward moves are marked with the score 500 (see Fig. 21). In the return climb the previously constructed tree is erased from memory.

Although in returning to the initial position the program found no satisfactory defense, its knowledge of the position has been considerably broadened: trajectories have been formed in the field connected with the Bishop's trajectory f2-g3-e5. The search for a move essentially begins anew, but now in the context of all the trajectories so far discovered. As before, the moves 1. ...(Ng7-h5 and 1. ...Ng7-f5 are studied, but in a different order, since the former has a higher priority as a move in a forking trajectory (the two trajectories g7-h5-g3 and g7-h5-f4 go through h5).

In the end, finding nothing new, the program returns to the move 1. ...Ng7-e6 in the blockade trajectory g7-e6-f4. White produces 2. Bf2-g3; Black blocks with 2. Ne6-f4. Here the program finds the trajectories g3-f4 for the Bishop and e5-f4 for the Pawn. To include these in the play, no backtracking is required. The exchange 3. Bg3:f4 e5:f4 is regarded as profitable in the first instance. The variation ends here, since the Bishop, which is the basis (the stem piece) of the field, has been lost and the exchange is complete. The score is zero, since the material lost on both sides is equal in value. We again backtrack up the variation with the assigned score.

Since the zero score for the move 2. Bf2-g3 does not suit White, the program substitutes 2. Bf2-d4 on the trajectory f2-d4-e5. We note, to the point, that now the Black Knight is on e6. After the move of the Bishop to d4 the program finds the trajectories e5-d4 for the Pawn and e6-d4 for the Knight, and these are at once included in the play. Priority is given to the Pawn capture, as being more profitable. The variation is ended, since the stem piece (the Bishop) is gone; the score is -3. We backtrack with this value, and in scoring the move 1. ...Ng7-e6 the program produces a minimax score and chooses the larger (more profitable for White) of the scores 0 and -3.

As a result of the search, the program has formed in memory a field of play. The search it has conducted cannot be considered a complete analysis of the initial position. The fact is that both the scores of the variations and the mathematical models formed by the program are provisional, since only one field has been formed and included in the play. Therefore the example presented here illustrates only the algorithm for forming a field.

1.4. Formation of a Field

By forming (in the initial position only) a stem trajectory, we form a field in the process of searching for a move and subsequently carry out the search itself by moving pieces along the trajectories of the field that have been constructed at the given moment. It is postulated that a trajectory for an arbitrary piece, from an a_0-square to an a_f-square will be constructed only if

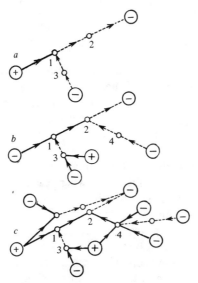

Figure 22   Stages in the formation of a field.

an enemy piece appears on the a_f-square during the search. Therefore, if a piece does not appear on some segment of a trajectory during the search, that segment will not be attacked and no higher-order denial trajectories will appear on it.

The proposal to connect the basic operations of Botvinnik's algorithm with the search for moves, and, in particular, the method for forming a field during the search process, was made by the author of this Appendix. It led to a decrease in the amount of information characterizing the model and also to a significant reduction of the search tree.

Some stages in the formation of a field are illustrated in Fig. 22 in which the dashes indicate the segments of the trajectories on which the search produced no attached trajectories of pieces. The procedure in Fig. 20 for searching out denial trajectories (considered below) will construct a field.

We should note that in using the term "trajectory" we have in mind a sheaf of trajectories, since the operation of defining a trajectory for a piece from some square to some other square, with a limited number of displacements (i.e., a trajectory of limited length), yields in the general case not a single trajectory but several [3], containing more information, especially if the length of the trajectory exceeds two displacements on an empty board.

1.5. Storage of Information about Sheaves of Trajectories in Computer Memory. Linked Lists

Before we can use the information we must solve the problem of storing it. Information on sheaves of trajectories is needed for the entire functioning of the program, since the formation of fields depends on it. What pieces of data must be saved? We decided to save in memory, not the sheaf itself (lists of squares in all its trajectories), but rather information that the sheaf exists and is of a given type, namely: the number of any piece (a uniquely assigned integer characterizing the piece) having a trajectory in the sheaf, the a_f-square of the sheaf, the length of the shortest trajectory in it, the type of the trajectory (blockade or control), and the parameters connecting this sheaf with others and with the higher levels of the control system. A complete list of the parameters is given in Fig. 23. In this subsection we describe the first four parameters; the others will be described later.

A sheaf is characterized by a list of parameters. The information in it must be bound to that square on the board where the piece was placed when the trajectory was first found, i.e., on an a_0-square. The sheaf is uniquely established by the first four parameters in the list, by means of the subroutine that obtains the sheaves of trajectories [3].

The data on the sheaves could be stored as a multi-dimensional array. We must remember, however, that each of the parameters is a two-digit decimal number (in the general case), and the linear dimension of the array is of the

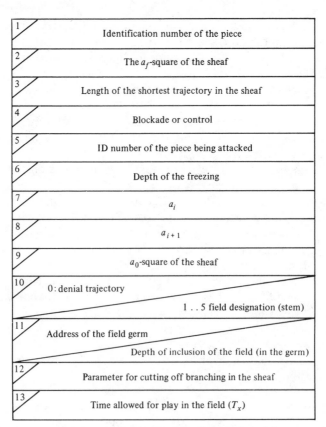

Figure 23   Structure of the standard cell of the linked chain (the trace of a sheaf).

order of 10^{26} (10^{13}), and this is obviously unacceptable. Therefore PIONEER saves the information in the form of a linked list (Fig. 24). The memory devoted to storage of a set of sheaves is divided into cells. Figure 20 displays 10 such cells. Each stores a parameter list and the address of the cell linked to it (shown in the upper right-hand corner of the cell). In addition, we reserve memory for two 8×8 arrays, in which the numbering of the elements corresponds to the numbering of the squares on a chessboard.

The information for all sheaves passing through an a-square is attached to the a-squares of these 8×8 arrays. This is done in the following way: in an a-square of the array I we write the address of the cell containing information on the first sheaf passing through the a-square, and then all cells with information on the sheaves related to that square are linked in a chain whose first element is the initial square. The address of the succeeding element in the chain is written in the corresponding a-square in array II. All cells in which, at a given moment, there is no information on these sheaves are also linked in a chain.

A system of this kind allows us to erase from some cells a certain amount of information that is not needed, either in the search process or after the piece has made a move in the play, and to use the freed space for new information. This is accomplished by linking the free cells in a list.

If, during a search, a piece visits an a-square, the information about all the sheaves of this piece that pass through the square is attached to the square. If the piece again appears on the square during the search, in an alternate variation, our information storage system allows us to restore all the sheaves of trajectories of the piece that relate to the given square and were constructed in the earlier search. This is done by sequential inspection of the linked list of cells associated with the square and by a search for the cells we need as computed from the parameters of the sheaves, and finally by execution of a subroutine for obtaining the sheaves of trajectories with

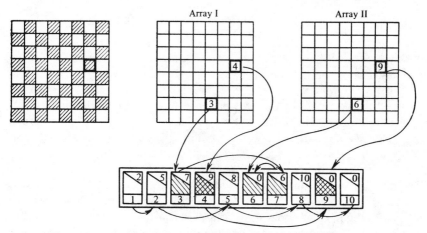

Figure 24   Storage of a set of sheaves in the computer memory.

these parameters. This operation is called *unravelling the sheaves*. A cell in a linked list attached to a given square and containing information about a sheaf is called the *trace of the sheaf* at the square.

1.6. Information on the Trajectories of a Field

Information on a trajectory of a piece in a field is connected with the *a*-squares of the trajectory, i.e., the squares on the chess board (an 8×8 array) which the piece has visited during a search (cf. Section A1.1.5). Thus, for every square in the 8×8 array, there exists a related linked list of cells, each containing information about one of the sheaves of trajectories passing through the square in question. Often the lists attached to various squares turn out to be empty. The standard structure of a cell in the linked list (trace of a sheaf) is shown in Fig. 23. The cell contains in particular the depth of the search tree at the node where, during the backtracking ascent, the sheaf must be unfrozen if it was frozen during the descent (see below). Moreover, if the given sheaf is a sheaf of blockade trajectories focussed on an *a*-square belonging to an attack trajectory and lying in an interval $b \subset (i, i + 1)$, the ends of the interval are written in the cell.

One of the entries in the cell is the address of the germ of the field, which contains information on the field to which the given sheaf pertains. The germ of a field is the trace of the stem sheaf of the field on the a_0-square of the sheaf. The address of this trace in the linked list is a parameter that uniquely characterizes the field. If the trace of a sheaf is itself a germ of a field, the place reserved for it under the address of the trace is used for another parameter—the depth of the tree at the point where the field was included. (Questions concerning the inclusion of fields in search trees are not considered in this Appendix.)

1.7. The Parameter T_x

A field is formed according to the following principle. The side marked $(+)$ may move only an a_0-piece in a stem trajectory, and other $(+)$ pieces are included in the field if they are on squares where they are ready to capture. The $(-)$ side includes in the field only those pieces that have time to take part in the play. The limiting time is defined by the variable horizon H_x—the time measured in half-moves allowed to the $(-)$ pieces for arrival in the battle on *a*- and *b*-squares of the stem trajectory [3].

Immediate use of the quantity H_x in the search for denial trajectories is inconvenient. We introduce instead an auxiliary parameter T_x which has the same dimensions as the length of a trajectory, i.e., a number of displacements. Changing with each move of a piece on the trajectories of the field (see below), T_x permits or prohibits the displacement of a given piece. It is also necessary for computing the maximum length of a trajectory of a $(-)$ piece that may be included in a field.

1.8. Computing the Length of a Denial Trajectory

Suppose that during the search some piece makes the move X-Y. We determine the trajectories of all the pieces that are on the board after the move X-Y and lead to the square Y. If a piece on square Z has a trajectory leading to Y, we say that there is a *sighting of Y from Z*, or that *Y can be seen from Z*. To register a sighting we need to know the maximum length (A_{max}) of the trajectories we are seeking.

Also, in order to preserve the information on a new trajectory of a piece from Z to Y, we need to fill in the trace cell, which is then linked to the list attached to the square Z in the 8×8 array. The parameter T_x is the most complex of those in the parameter list of a trace.

We must determine two quantities: A_{max} and T_x. These depend on the nature of the trajectories we are looking for.

If we are computing the stem trajectory of a field, $A_{max} = \text{Ent}[(H_L + 1)/2]$, where H_L is the horizon, in half-moves; $T_{max} = 1$.

If we are seeking a denial trajectory, both A_{max} and the initial value of T_x depend on which trajectory the desired one is attached to, i.e., in which trajectory of the field the move X-Y was made. To determine this, we study the traces of trajectories at X and Y. From these traces we can extract the value of T_x, say T, for the trajectory of the piece that made the move in question, and also A, the length of the remaining portion of the trajectory.

Let us look at all possible cases.

(1) The piece moved on a stem trajectory. Then the length A_{max} of the desired denial trajectory does not exceed T (see below, on the change in T_x for motion on a stem trajectory): $A_{max} \leq T$. We assign the value 1 to the parameter T_x.

(2) The piece moved on a denial trajectory:
  (a) to the terminal square of the trajectory; then $A_{max} = T - 1$, $T_x = T$;
  (b) not to the terminal square, and on the remaining portion of the trajectory the shortest path in the trajectories of the sheaf is of length A; then $A_{max} = T - A$, $T_x = A_{max} + 1$.

If in the cases (1) and (2) it is a b-square that is under attack, the values of A_{max} and T_x are computed in the same way, and then decreased by 1. For the cases considered here, in computing the maximum length of denial trajectories (A_{max}), we have in mind the trajectories of the $(-)$ side. It is taken for granted that for trajectories of the $(+)$ side in the field, with the exception of the stem trajectory, the length is $A_{max} \leq 1$.

Thus, in a field of play, among the $(+)$ pieces only the stem piece may move (if there is no capture); controlling $(+)$ pieces must be in ambush, i.e., their trajectories are of length 1. The limiting length of the denial trajectories of ($-$) pieces is defined by the parameter T_x, and therefore ($-$) pieces are included in the field only if they have time to take part in the play.

1.9. Correction of the Sighting Method

The search for denial trajectories by the direct sighting method, as compared to other methods, results in a significant decrease in the amount of information that must be stored in memory. We are speaking, for instance, of an a priori construction of a complete field. However, the sighting method often leaves both trajectories and whole fields incomplete. This happened in our earlier example of the formation of fields of play. A field is formed only in those places where pieces really have visited during the search, where a battle has occurred. The areas not so visited do not become covered with trajectories and therefore the computer memory is not burdened with useless information. This may broaden the search for moves.

However, the sighting method presupposes that at the moment of sighting, after the move X-Y, the square Z from which we seek a trajectory leading to Y should be occupied by a piece; but this is not at all prerequisite for the existence of a trajectory. Let us suppose that in the course of the play in the field some piece visited the square Z, and, having a trajectory from Z to Y, could arrive in time to take part in the battle at Y. But at the moment of sighting it has already left Z, and the corresponding trajectory is not found. Thus in the sighting method we lose some trajectories.

This shortcoming of the sighting method was discovered only in the course of experiments with PIONEER. We can offset it by correcting the method. Suppose that a piece has made the move X-Y along some trajectory in a field. Then we can determine the depth of inclusion of this field in the search tree, since it is written in the germ of the field (cf. Section A1.1.6). On obtaining control, the search procedure for new denial trajectories generates a reversal in the current variation, wherein the ascent up the tree is implemented to a depth equal to the depth of inclusion of the given field. At the time of the ascent, two operations are executed simultaneously with a backward move: the position of the piece is changed and T_x is recalculated for the trajectories passing through the field X-Y (see below). In the position at the end of the backtrack, the sighting method will produce all denial trajectories leading to the square Y (and all the blockade trajectories for the segment X-Y); the quantities A_{max} and T_x for the new trajectories are determined as described above.

The procedure generates a descent along the same branch of the tree that we have just climbed. At every move X_1-Y_1 in the descent we compute new trajectories for the piece making this move, that is, from the square Y_1 to the square Y [on the segment (X-Y)]. The ascent along the branch of the search tree and subsequent reversal corrects the shortcoming of the sighting method.

1.10. The Pseudosearch

When we find a new denial trajectory of length greater than one displacement, the motion of a piece along it makes no sense in the given variation. (The length of a blockade trajectory is arbitrary.) This happens because what we have is either a blockade trajectory ending in some b-square of the attacked piece, and that piece has already bypassed the square, or it is a trajectory for the control of some a-square and the attacked piece is already on that square, so that the control is too late. Because the play cannot include trajectories that have been found too late, further pursuit of the variation is senseless. Therefore we must stop the descent and back up until we reach a point at which the new trajectories can be included in the play. The subtree consisting of the branches issuing from those nodes of the current variation that we encounter during the backtracking must be erased from memory, and the corresponding subtree of the search for a move must be tagged as provisional.

We therefore call the ascent up a branch of the tree (corresponding to the current variation in the search) a *pseudosearch*, which is undertaken only to include new trajectories in the play.

1.11. The Extent of the Climb Up a Branch in the Pseudosearch

Let us suppose that at some point in the search we have found new trajectories for various pieces. Let us determine which of these pieces has stayed longest on its a_0-square in the current variation. One would think that the highest node in the tree at which this piece is still on its a_0-square is the extent of the climb, but we must verify the existence at this point, of the field to which the trajectory belongs, since we may have climbed beyond the depth of inclusion of the field. Thus, we must compare the two numbers: the extent of the climb to the highest node at which the oldest piece still stands on its a_0-square, and the depth of inclusion of the given field; then we take the larger as the extent of the climb to the node at which the new trajectories of the field are to be included in the play.

We note that all the depths in the initial position are equal to zero and increase in the course of the variation. Further, we take into account the fact that in the general case the search proceeds in more than one field, and the trajectories or segments of them belong simultaneously to several fields. We have shown how to determine the extent of the backtracking for each of these fields. After calculating these extents, we have several numbers; the smallest is the final extent of the climb in the pseudosearch. The pseudo-

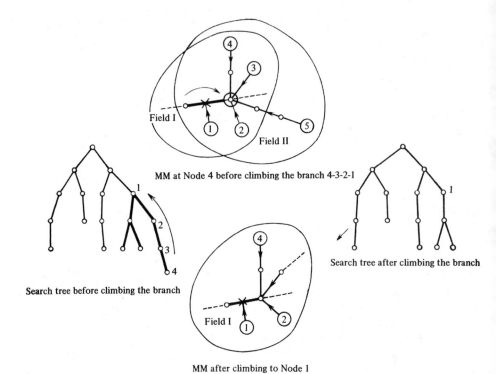

MM at Node 4 before climbing the branch 4-3-2-1

Search tree before climbing the branch

Search tree after climbing the branch

MM after climbing to Node 1

Figure 25   The pseudosearch procedure.

search procedure is illustrated in Fig. 25. The field designated as II has a very short life, and perishes during the climb; all its trajectories, including the newly found, are erased from the computer memory.

1.12. Freezing Trajectories Because T_x Is Too Small

Let us now see how the parameter T_x is recalculated for every $(-)$ piece in a field after each move in the field. It is calculated during the descent of the tree, by a procedure for computing it after a step forward, and during the climb, by a procedure for computing it after a step backward (cf. Fig. 20). If the stem piece moves on its stem trajectory, its own parameter T_x is increased by 1, and on the denial trajectories, T_x is decreased by 1. For the stem trajectory T_x serves only to determine the maximum length of newly computed first-order denial trajectories. If during the search a move is made along a denial trajectory belonging to a field, T_x is increased by 1 on the trajectories of higher-order denials connected with the trajectory on which the move was made. On the remaining trajectories of the field, T_x does not change after such a move.

Suppose that a $(-)$ piece is on the square i of a denial trajectory belonging to a field, that the shortest path from i to the terminal square of its trajectory has length A (displacements), and that the time allowed for play in the field on this trajectory is defined by the parameter T_x. The rules for play in the field are such that the given piece is allowed to make the move $i - i + 1$ if (1) the move results in a capture or (2) $A \le T_x$ and the trajectory has not been frozen by other criteria (see below). Thus, if $T_x < A$, the trajectory is *frozen by an insufficient value of the parameter T_x*.

In backtracking up the tree, T_x changes in a similar fashion, except that the changes made on the way down are reversed: if a given move during the descent increased the value of T_x by 1, the same move when made during the backtracking decreases it by 1; if in the downward path the move decreased T_x, then on the backward path it increases it by the same amount. If the move left T_x unchanged on the way down, it leaves it unchanged on the way up. Thus, in the process of search and minimax, having descended from a node of the search tree, in returning to it we provide for the restoration of the value of T_x for all trajectories of the field.

Figure 26 presents an example of the calculation of T_x for various denial trajectories. The value of T_x in the initial position of the pieces is obtained in the following way (for simplicity, consider only piece 4): In accordance with the rules for completing the trace of a stem sheaf in the initial position, we have $T_x = 1$ for the stem piece. Let us suppose that the denial trajectory has not yet been constructed. When (in the search process) we move the stem piece, the value of its T_x is increased by 1 at each move. After the move 3-4, when it is on the a-square 4, its T_x value is 4. Let us find a trajectory for piece 4 with a length not exceeding four displacements. For this piece, $T_x = 1$ (see Section A1.1.8). When control passes to the pseudosearch procedure, the climb back to the initial position of the stem piece begins. With each backward move of the stem piece, the value of T_x for piece 4

Move		Value of T_x				
		Piece				
		1	2	3	4	5
Initial position of the piece		3	4	3	4	2
1	11-10	3	4	3	4	2
	1-2	2	3	2	3	①
1	10-9	2	3	2	3	1
	2-3	1	2	①	2	0
2	12-8	2	2	2	2	1
	3-4	1	1	1	①	0

The instants at which the corresponding pieces can no longer move in the field: $A = 2$, $T_x = 1$

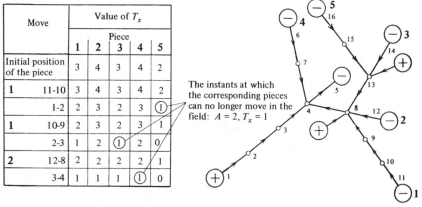

Figure 26　Simultaneous search and computation of T_x.

increases by 1, and therefore, in the initial position $T_x = 4$. Let us note here that if a trajectory exists at some node of the tree, but remains undefined for lack of a sighting from that node, and is first defined at a node further down the tree, then its initial value of T_x (from Section A1.1.8) is such that when the return journey up the tree is made, the changes in T_x bring its value at the first node up to what it would have been if the trajectory in question had been discovered there on the way down.

1.13. Recalculation of T_x on Denial Trajectories

Here we describe in detail the algorithm for recalculating T_x when a piece moves on a denial trajectory. In accordance with our rule, we must find the highest-order denial trajectory that belongs to the field and connects with the trajectory of the given piece. Then we increase the value of its T_x by 1.

To this end we choose an arbitrary trajectory in the field, then transfer to the trajectory it connects with, transfer again to the one the latter connects with, and so on. Then, in going from one trajectory to another (meanwhile decreasing the order of denial at each step) we may come at last to the stem trajectory of the field. If so, we do not change the T_x of the initial trajectory. Then we go from trajectory to trajectory in another variation, and if we come to the trajectory in which the given move was made, we change T_x for the initial trajectory. We repeat this procedure for every trajectory in a denial field.

1.14. Freezing a Trajectory for Lack of a Connection to an Active One

There is an unfreezing procedure, executed during the backtracking process (cf. Fig. 20), which corresponds to the freezing of a trajectory during the descent of the tree. It is implemented as follows:

When a piece moves along the trajectories of a sheaf, there appear at each step portions of trajectories, or even whole trajectories, on which it has not yet set foot during the variations developed in the search. Trajectories connected with these portions must be excluded from the play, together with higher-order denial trajectories associated with them. We call this operation *freezing for lack of a connection with an active trajectory*.

Figure 27 illustrates the freezing procedure for the move 1-2 of piece 1. The following segments and trajectories are frozen: 1-2, 1-3-5-6, 14-9-5, 13-3. In backtracking up the variation, the corresponding trajectories must be re-included in the play and used in other variations of the search. This operation is called *unfreezing*. The exclusion of trajectories at a node of the search tree during a descent, and the subsequent re-inclusion at the given node during the backtracking, are called *freezing for lack of a connection with an active trajectory* and *unfreezing by resurrection of the connection*, respectively. The procedure works as follows:

1.15. Criteria for Freezing

We consider an *a*-square common to several sheaves of trajectories of a given piece. We say it is totally frozen if all the trajectories of these sheaves that pass through the square in question are frozen. For example, in Fig. 27 the squares 1, 3, and 5 are totally frozen for piece 1, and square 5 is totally frozen for piece 4. The control trajectories for an *a*-square freeze only if the square itself is totally frozen. For instance, the trajectories 13-3 of piece 2 and 14-9-5 of piece 5 are frozen. A blockade trajectory focussed on a *b*-square is frozen on the segment $b \subset (a_i, a_{i+1})$ only if one of the endpoints a_i or a_{i+1} is totally frozen. The trajectory 12-11-10 of piece 3 in Fig. 27 is not frozen, since the endpoints of the interval (9-14) are not totally frozen. As we have noted, the endpoints of the interval (a_i, a_{i+1}) for a trajectory blockading a *b*-square are entered in the memory-stored list of data concerning the trajectory. (See Fig. 23.)

These are the criteria for freezing high-order denial trajectories connected with a given trajectory.

When an *a*-square belonging to a sheaf is frozen, the current depth in the search tree is written into the trace. When the backtracking along the branch reaches the same node, all sheaves with the given depth are unfrozen and the depth of freezing is set equal to zero. (With a zero depth no sheaf is frozen.)

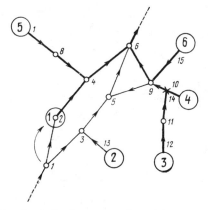

Figure 27   Freezing for lack of a connection with an operational trajectory.

1.16. The Algorithm for Freezing

Suppose the move X-Y has been made in the search process; let us analyze the freezing procedure (Fig. 28). First we determine which sheaves contain the trajectories on which the move was made. In a special 8×8 array we mark the squares belonging to these trajectories and pay attention to those and only those a- and b-squares which the piece has not yet reached. The information about all the chosen trajectories is contained in the corresponding cells in the linked list attached to the square Y. Control is then passed to the freezing procedure. We search all squares of the board except those marked in the special 8×8 array, seeking information on any not yet frozen trajectories of the given piece that pass through the square X. If at some square Z we find information on such a trajectory, i.e., a trace with zero in slot 6, the a-squares X and Z of this trajectory are frozen. This procedure freezes the trajectories of the given piece [Fig. 28(b)]. (In Figs. 28(a)–28(d), the frozen sheaves are shown as shaded areas.)

Now we must freeze the higher-order denial sheaves. We again inspect all squares of the 8×8 array for information on non-frozen sheaves for every piece except the one that made the move that started the procedure. If at some square V we find information on such a sheaf, we extract from the corresponding cell the information it contains on lower-order denial sheaves with which the newly discovered sheaf is connected. Then we apply the criteria given above for freezing the higher-order denial sheaves. If one of them is satisfied, the a-square V of the given sheaf is frozen and the inspection of the 8×8 array continues. After the first inspection we have

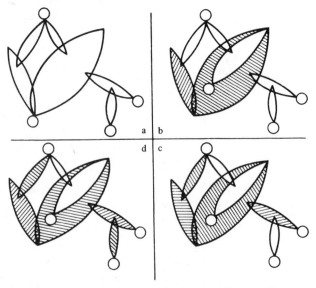

Figure 28   Sequence of operations in freezing sheaves of trajectories.

frozen the sheaves connected with the frozen trajectories of the piece that made the initial move. [Fig. 28(b)].

We denote the set of these sheaves by $\{F\}_1$. We again inspect the 8×8 array and execute all the operations performed in the first inspection. The result is the set $\{F\}_2$ of frozen sheaves connected with the sheaves in $\{F\}_1$. We continue this process until after some nth inspection we find that no sheaf was frozen, i.e., $\{F\}_n$ is empty. (For the case depicted in Fig. 28, $n = 3$.) At this point the freezing procedure ends.

1.17. The Freezing Procedure on Leaving an a_0-square While Backtracking

Freezing on exit from an a_0-square in a backtrack corresponds, on return to the a_0-square, to the unfreezing which was executed on the way down (Fig. 20).

During a backtrack a piece may leave the a_0-square of its trajectory, because many trajectories do not start from the initial position but rather at some depth in the search when the pieces have already left their initial squares. Therefore in a backtrack we may find a non-frozen trajectory of a piece that is not even standing on any square of its own trajectory. If the stem piece of the field leaves its a_0-square, all the trajectories of the field are scrubbed; for a non-stem piece it is more advantageous to freeze the trajectory. This freezing, however, is different in nature from those we have considered so far. First, it occurs in a backtrack, and the unfreezing occurs in the descent (if the piece appears on its a_0-square). Second, entire sheaves are frozen, not merely separate portions of trajectories.

The procedure is shown in Fig. 29. In a backtrack, after the inverse move Y-X, we inspect the whole 8×8 array for sheaves having Y as their a_0-square. We can do this because information about the a_0 of a sheaf is given in the corresponding cell of the linked list (Fig. 23). We write the provisional number 99 into the "depth of freezing" slot in the traces of the sheaves that we have discovered. We use a slightly different procedure for freezing higher-order denial sheaves; it is like the one we used for the loss of connection with an active trajectory (cf. Section A1.1.16).

If during the descent a piece arrives on the a_0-square of a sheaf that is frozen with the 99 mark, that sheaf and all sheaves connected with it are unfrozen. The corresponding procedure, which is like that for freezing on leaving an a_0-square, generates a search for all sheaves frozen with a 99 tag and having the given square as their a_0-square. Those found are unfrozen together with their associated higher-order denial sheaves.

The unfreezing of trajectories because a piece arrives on an a_0-square during the descent does not imply their inclusion in the search. Rather, the inclusion is determined by all three types of freezing and, in particular, by the variable horizon H_x (in the program, however, this is controlled by the

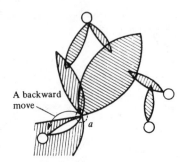

Figure 29   Freezing sheaves when a piece leaves its a_0-square while climbing a branch of the search tree (the frozen sheaves are shown as shaded areas).

parameter T_x). Even when some trajectories are frozen, T_x is continuously recalculated for all occupied squares in these trajectories. Therefore when a piece returns to the a_0-square of its trajectory during the descent, we can at once decide whether to include its trajectories by referring to the T_x value assigned to that square.

1.18. Trajectory and Field

In this subsection we consider the interaction of the two lowest-level control systems: trajectories and fields. These are hierarchically ordered: under system control, i.e., during the search for moves, pieces considered as elements may move only along the trajectories constructed for them at the time. Other moves permitted by the rules of chess do not exist for them in the current model of the game. Thus the moves of the pieces are subordinate to the rule of the trajectory, which represents the first level of the control system.

A field of play consists of a collection of trajectories and pieces assembled in two camps, labeled $(+)$ and $(-)$ as representatives of color, in a battle for:

(1) some piece which is the goal of the play in the given field, or
(2) some square indirectly connected with an attack on a piece.

The field, as the second level in the control system, constantly influences the moves of pieces in their trajectories. This influence is exerted by three different procedures that either prohibit or permit such movement. These procedures are the three different types of freezing:

(a) for shortage of time allowed the piece for play in the field, as determined by the parameter T_x;
(b) for lack of connection to an active trajectory;
(c) when a piece leaves the a_0-square of its trajectory in the backtrack.

The field intervenes in the operations of its subordinate level, the trajectory, by organizing and influencing the movement of pieces. This intervention is justified by the fact that the goal of play in the field at any given moment may differ from the goal of the piece in a trajectory of the field; for instance in a denial trajectory the goal may be the control of the a_0-square of the trajectory, with the immediate goal of capture by ambushing the piece that is to be controlled.

The type of a trajectory is assigned to it by the higher-level control system, and the type determines its goal. For the stem trajectory of an attack field the goal is to capture the a_f-piece. For a control trajectory (a denial of the stem trajectory of a denial field) the goal is to annihilate the piece to be controlled by waiting in ambush on the a_{f-1}-square. A blockade trajectory has no goal of its own; it is wholly subordinate to the goal of the field in which the piece to be blockaded plays.

Thus, the hierarchy of control levels is accompanied by a hierarchy of the corresponding goals of play. The field itself is controlled by the third level of the system—the mathematical model. This control is again justified by the differences in the goals of the game at the second and third levels.

The field reacts to this control during the search for a move by influencing its subordinate levels. In directing the search, the field is formed by the search itself, i.e., the denial trajectories are constructed. Thus there is a feedback process connecting the first and second levels, which changes the structure of the control system itself. This change directs the formation of fields and results in an optimal variation of the system control.

2. The Choice of Moves in an Ensemble of Fields (The Mathematical Model)

Now let us see how the search for moves proceeds under the control of an ensemble of fields. In chess the goal of each side is the mate of the opponent's King. In our model, it is replaced by an intermediate goal—winning material. Each side, pursuing this goal, chooses its own optimal strategy, i.e., its best variation. The set of all strategies forms the search tree. The minimax choice of the optimal variation in the search tree amounts to the choice of the optimal strategy in the control system. We now examine PIONEER's formation of the search tree.

2.1. The Goal of the Game and the Ending of a Variation

One of the most important tasks in the development of a chess-playing computer program is the establishment of its criterion for scoring a variation and ending it. Let us define a quiescent position as one in which neither captures nor checks/responses are in view. As a rule, the existing programs for playing chess continue a variation to a predetermined depth or to a quiescent position, whichever comes later. The end positions in a variation are scored by a linear function that computes the material balance over the whole board and takes account of many positional factors. Such an approach to the development of the search tree does not find good moves in the initial position, nor is its method for ending a variation supported by the way a chess master plays.

In the algorithm we are now discussing, and therefore in PIONEER, the ending of a variation is controlled by the goal of the play. A variation is ended if its goal is either reached or found to be unreachable. Since we are contemplating a multi-level control system, and the several levels have different goals, the criterion for breaking off a variation, i.e., stopping the play, is differently formulated for each of the levels.

2.2. The Criteria for Breaking Off a Variation in the Search

The goal of play in a field is the capture of the a_f-piece, i.e., the piece attacked by the stem piece in the stem trajectory. The value of the a_f-piece is the gain striven for by the $(+)$ side over the course of play in the field. The $(-)$ side opposes this gain. His goal of play is to lose nothing, if possible, or at least lose less than the value of the a_f-piece. Thus the criteria for terminating a variation are in essence the expectations by either side of reaching their goals by continuing the variation. A variation of play in a field is broken off if one of the following conditions is satisfied:

(1) the a_0-piece is captured or is frustrated by an irremovable blockade;
(2) the a_f-piece is captured or leaves the a_f-square;
(3) the loss of material (m_T) is greater than the value that can be gained, from the viewpoint of the $(+)$ side, or preserved, from the viewpoint of the $(-)$ side. That is, $-cm_T \geq m_f$, where m_f is the value of the a_f-piece. If in the given position the move belongs to White, $c = 1$; if the move is Black's, $c = -1$.

We now consider the mathematical model (MM), consisting of several attack fields included in the play. There will be a number of a_0- and a_f-pieces, possibly of different colors. To formulate the criterion for breaking off a variation in the MM, we must calculate the sums Σm_w and Σm_b of the values of the White and Black a_f-pieces in the several fields. Then the criterion of expectation becomes:

(4) $-cm_T \geq \Sigma m_w + \Sigma m_b$, where c is the same as in condition (3) above.

Thus, in the search of a variation in the MM, play is cut off in those fields where either condition (1) or (2) is fulfilled, and is continued in the other fields. If condition (4) is satisfied, the variation is terminated in all the fields. When the highest level in a control system abandons hope of reaching its goal, play in all lower levels must stop, even though the local goals may still be attainable. Condition (3) is not tested for the individual fields.

Thus the criteria for ending a variation must be tested (and consequently the quantities in condition (4) must be computed) after every move during the search through variations in the MM.

2.3. Testing the Criteria for Ending a Variation

The satisfaction of criteria (1) and (2) is guaranteed by the freezing and unfreezing procedures (cf. Section A1.1.14). If either of these two criteria is satisfied in some field belonging to the ensemble, the trajectories of the stem sheaf are frozen, and therefore so are the denial trajectories; this means that the search in the given field is suspended and goes on in non-frozen fields.

Let us now see how we compute the sums Σm_w and Σm_b for criterion (4). Consider a stem piece appearing in the ensemble of fields, for instance piece 1 in Fig. 30. Suppose that in scrutinizing our information on the stem sheaves of trajectories connected with the square on which piece 1 stands, we find that its attacks on pieces 3, 4, and 5 have not been frozen. Using a procedure for constructing trajectories [3], we find those belonging to these sheaves. Consider one such trajectory, e.g., 1-2-4-5-7. Inspecting its a- and b-squares in sequence, we note that piece 3 of the $(-)$ side is standing on the a-square 4, and that there exists a non-frozen attack trajectory belonging to piece 1, namely, 1-2-4. Accordingly, in the movement of piece 1 along the stem trajectory 1-2-4-5-7 the maximum gain is $m_3 + m_5$, where m_3 and m_5 are the respective values of pieces 3 and 5. Turning our attention to the other trajectories in the stem sheaves of piece 1, we find that the maximum gain for 1-2-3-5-7 is m_5, for 1-2-4-5-6 it is $m_3 + m_4$, and for 1-2-3-5-6 it is

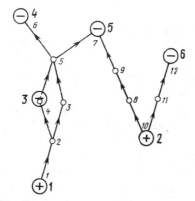

Figure 30   Stem trajectories in an ensemble of fields.

m_4. Thus the maximum possible gain for fields with the stem piece 1 is equal to $\max(m_3 + m_5, m_5, m_3 + m_4, m_4)$. Suppose it is realized on the trajectory 1-2-4-5-7, and is therefore equal to $m_3 + m_5$. In our special 8×8 array we mark the squares containing pieces 3 and 5.

Now we pass to the next stem piece in the ensemble of fields, piece 2 in Fig. 30. Proceeding as before, we find that the maximum gain in fields with the stem piece 2 is equal to $\max(m_5, m_6)$. We note, however, that the square containing piece 5 is already entered in our special 8×8 array; piece 5 is therefore excluded from consideration; our maximum is then simply m_6. In the ensemble of fields in our example (Fig. 30), we have $m(-) = (m_3 + m_5) + m_6$, $m(+) = 0$.

The running value m_T is also computed after each move in the search process, as $m_T = (M_b - M_w) + m_0$, where M_w (M_b) is the sum of the values of all White (Black) pieces removed from the board in the course of the current variation, calculated from the initial position to the current move; m_0 is the material relationship in the initial position.

When a variation is broken off, the value of m_T in the final position is its final score m_f. Thus the model under discussion here is characterized by the absence of a static scoring function. The function m_f is evaluated only at selected nodes, namely at the terminal node of a variation broken off by the criteria listed above. Thus one of the fundamental tasks in this model is to form the domain of definition of the scoring function m_f. We must again emphasize the fact that the scoring function here is not merely an assessment of the final position of a variation; it is also an assessment of the variation itself in the search, as a strategy in the control system, i.e., the extent to which the goal of the play is being reached by the current strategy. If it were isolated from the variation, i.e., treated as merely the material balance in the given position, it would have no sense.

We note that PIONEER's scoring function contains a positional component closely connected with the MM; however, a discussion of it is beyond the scope of this exposition.

2.4. Pruning Branches by the Minimax Principle. On the Branch-and-Bound Method

We construct a deep and narrow tree by an a priori rejection of certain moves at each position, i.e., moves not on a trajectory. In the established terminology, this is known as *pruning in the forward direction*. In backtracking up the tree in connection with the minimax principle, however, it may happen at some node that a further descent from it will not change the outcome of the minimax procedure. Then we may prune the branches at this node and resume the backtracking. This means excluding certain moves in the corresponding position; this pruning of branches while minimaxing is called *pruning in the backward direction*.

Let us see how minimax pruning takes place in our current model of the game. We suppose that a search is under way in an ensemble of fields. Suppose that during the minimax procedure we find at some White node a final score for our variation which is so high that the opposing side will have cut off the path to the node somewhere up the tree. Then the node is in practice unreachable and no branches emanating from it need to be considered.

Let us look at two methods for pruning used in our model.

The first is based on the following argument: Suppose that the White node N envisioned in the preceding paragraph is a branch point of the tree and is a minimax point. Therefore current scores will exist at certain nodes. Then after backtracking to N and determining its current score m_1, we must examine each of the Black nodes higher up in the current branch to see whether at any of them the current score exists and satisfies $m_2 \leq m_1$. If we find any, no branches are constructed from the node N.

This argument is the basis for the branch-and-bound method, sometimes called the α-β-cutoff method, used in almost all chess-playing programs and in some other search problems. Clearly, the sequence in which moves are considered is of great importance. It has been shown [8] that given the theoretically best search sequence, in a tree of dimension a^n (a is the fixed number of moves in each position and n is the depth of search in half-moves), the branch-and-bound method reduces the search to a magnitude of the order of $a^{(n+1)/2}$. In practice, in the existing chess-playing programs, the magnitude of the search to a fixed depth in the search tree greatly exceeds the theoretical minimum, reaching 10^5 half-moves for a depth of 5 half-moves. The tree grows in size by a factor of 7 with each additional half-move in the depth of search, at least in the middle game.

Programs with such a search tree place heavy demands on the speed of a computer. Even if this high price is paid, in the form of a very powerful computer, we do not get an essential improvement in the play, since it is impossible to reach master-level chess by adding one or two plies to the depth of all variations. Moreover, in search problems where in every situation the number of possibilities is much larger than the number of moves in a chess position ($A \gg a$), the practical value of a full-width search to a fixed depth is extremely small (even if the branch-and-bound method is used) because of the astronomical size of the search tree.

Let us now return to the definition of the model of the game. We shall not use the branch-and-bound method as such, since an establishment of the limits of change in the scoring function before beginning the search would distort the mathematical model. However, we do use the arguments given above which underlie the method.

During the search, the computer remembers the current depth of the highest node on the tree that has a current score. With every upward step, a current score is determined at some node; the program ascends the tree and compares scores only to the remembered current depth.

Let us now look at the second method for pruning branches, which is peculiar to our algorithm.

2.5. Pruning Branches by the Worst Case Method

We define two numbers at every node—the value of the goal and the value of the variation. Comparing these values, we know whether we must construct or prune any other branches issuing from the given node.

Since we are considering an ensemble of fields, the value of the goal may differ between Black and White, and so may the decision as to whether or not to continue; this depends on which side has the move at a given node. Suppose White has it. We introduce the notion of the worst case: this is the outcome, the score m_f, when one side proceeding from the given node in his current optimal variation in the ensemble of fields would gain nothing and lose as much as possible. For Black this is the loss of the values of all the a_f-pieces, i.e., $\sum m_b$. Thus, when a final score m_f appears at a White node, and

$$- m_f \geq m_T + \sum m_b,$$

other branches from the White node need not be formed, since the minimax procedure will in any case exclude this node from the optimal variation. Similarly, at a Black node the criterion for pruning is

$$- m_f \geq m_T + \sum m_w.$$

It is not difficult to test the criteria for pruning. After each move during the backtrack along a variation we calculate $\sum m_w$ and $\sum m_b$ over the ensemble of fields, by using the procedure already applied for testing the cutoff criteria.

It is worth noting that the need for pruning branches by a minimax process, and the necessity for including the corresponding procedure in our model, were both brought to light only during our experiments (with the search for moves in a field) under a working program—the first version of PIONEER.

The model contains another procedure for cutting off branches at a given node during a backtrack, which offers a significant saving in the number of moves to be inspected. This process (to be discussed in Section A1.2.10) deals with the elimination of fields and the pruning of branches in a sheaf.

2.6. Priority of Moves in a Search

The gain from using the minimax-pruning procedure, i.e., a significant shrinkage of the search tree, can be obtained only if during the descent along the tree we know which branches to form first at a given node in order to solve the pruning problem for the remaining branches. We set up the following system for assigning priorities to moves being considered for inclusion in the play in fields:

1. The fundamental goal in our model is to gain material, and therefore captures get first priority. The more gainful the capture, the higher the probability of cutoff; therefore the first priority goes to captures with the highest value of the captured piece, and to exchanges giving the highest material value gained on balance.

To implement this principle we need a quick recognition of the captures among the moves in the trajectories of the MM. (Note that a field always contains at least one capture trajectory.) The recognition can be achieved by a procedure called "moves not in the MM" or "total move generator" which constructs all legal moves of a piece, not only those in the trajectories of the MM. The basis of this procedure is an embedding of an 8×8 array in a corresponding 15×15 array [3] and the marking of squares in the 8×8 array that can be reached by a legal move from the starting square.

The procedure for finding the most profitable captures is as follows: Select a piece that (1) belongs to the side that is seeking a move at some node of the search tree, and (2) is in the field where the search is under way. Use the total move generator to determine all its captures. Select one of them. This yields a pair of pieces: the victim, with value m_1, and the attacker, with value m_2, together with the coordinates of the two pieces. Then, by a procedure described below, see whether this capture has already been found in the earlier portion of the search. If it has, go to the next capture; else assign the coordinates of the attacker and the victim to the variables X_1 and Y_1, respectively. Assign the value $300m_1$-m_2 to the variable D_{max}. Go to the next capture; redetermine m_1 and m_2; if $300m_1$-$m_2 > D_{max}$, store in memory the new values of X_1, Y_1, and D_{max}; else go to the next capture and repeat the process. In the end, the triplet X_1, Y_1, D_{max} will represent the most profitable capture for the selected piece, and the trajectory will be that corresponding to the endpoints X_1, Y_1.

2. In forming a field, i.e., constructing new denial trajectories, it is advantageous to include entire regions (collections of mutually related denial trajectories) since if the result of the combat in the field is decided too narrowly, other regions of the field will not be formed because the variation will be pruned or cut off by the minimax procedure. This means that the search and the MM will be narrower. Therefore the second priority goes to moves in trajectories recently included in the MM.

This principle may be implemented by attaching a distinguishing mark to newly constructed sheaves and, on returning along the variation in the pseudosearch, using these marks to decide which sheaves to select with first priority. As new trajectories are found, the marks on the old ones are to be erased.

3. We assume that the pseudosearch (cf. Section A1.1.10) has led to the construction of new denial trajectories. We backtrack to some node, erasing branches as we go. Then the search resumes from the given node with the inclusion of new trajectories in the play.

We note that the movement of a piece attacked during a pseudosearch must be identical to its movement in the preceding variation, which was preliminary and erased from memory. Therefore the inverse motion of an attacked piece is remembered in the traces attached to the *a*-squares occupied by the piece in its inverse trajectory. Then during the forward search the former motion may be repeated, since the moves of the given piece have priority in the marked squares.

4. Chess time, measured in plies (half-moves), plays an important role in our model, since a goal is attained most quickly by moving pieces over the shortest paths. Moves in such paths get fourth priority. Given two paths of equal length, preference goes to the forked paths, i.e., those for which a part is common to several trajectories; an example is provided by the paths 1-2-3-5-7 and 1-2-4-5-6 in Fig. 30.

We have not yet completely implemented this principle in our program, but we have programmed some particular cases. If the program adopts a decision on the movement of a piece in some sheaf of trajectories, priority goes to the shortest (forked) paths in the sheaf (see below). We expect to compare the forking of trajectories of various pieces in the next version of PIONEER.

Our experiments with PIONEER showed that the current system of priorities needs to be made more precise. It does not take into account the degree of importance of a field; this concerns the value of the goals, the distance between the attacking piece and the victim, and the practicability of the trajectory. With respect to practicability, we have in mind the following: Consider all the *a*-squares of the path of a piece; select all pieces one move away from a selected one of these squares. Determine the outcome of the optimal exchange on it for these pieces. Do this for each *a*-square, taking account of the fact that the original piece is moving in its trajectory and arrives on the selected square.

An *a*-square for which the exchange is favorable to the side owning the trajectory will be called a *practicable square*. A field with a stem trajectory on which all the *a*-squares are practicable will be called a *field of vulnerability*. Naturally, such fields should receive high priority; the highest priority goes to attack fields.

These principles will be taken into account in the next version (see also Appendix 2).

2.7. Analysis of the Trajectories of a Sheaf for Inclusion in the Search

We have noted that if the program selects a sheaf of trajectories for a piece, during the search for variations, first priority goes to the paths having the smallest number of moves, and if there are more than one, to the forked paths. The necessary information is in the trace attached to the square containing the given piece. We determine whether or not the sheaf is frozen.

If it is not, we use the procedure described in Section A1.1.5 to develop the sheaf, i.e., we construct all the trajectories explicitly on an empty board. Next we transfer the trajectories, one at a time, to the real board, find their a- and b-squares, and calculate their lengths. On this basis we determine whether there is time for the piece to move in these trajectories; we need only compare the current value of the parameter T_x for a given trajectory with its actual length (see Section A1.1.12).

If the trajectory is blocked by pieces of the same side, we determine whether it is possible and necessary to deblockade it. If the portion $[a_0 - a_1]$ is blocked, that portion, as a trajectory, does not appear in the search, but there is information on the necessity for a deblockade, contained in the information about the square containing the blocking piece.

Figure 31 displays as STEP1 the array of squares on the board that can be reached in one move from the initial position of the piece. STEP1 is shown symbolically as a broken ring, to represent the fact that certain moves of the given piece at the current node have already been made during the search of the variation. (The determination of these moves will be discussed later.) Now suppose that we have analyzed a trajectory in the sheaf and have found that with respect to all its parameters it may be included in the search. Then we mark its a_1-square in STEP1, which obviously contains it. We store the length (A) of this trajectory. Assume that some trajectories found later in the search may also be included and have the same length. We mark their a-squares in the array STEP1. This is shown in Figs. 31(a) and 31(b).

In going from one trajectory to another we may arrive at one whose length on the real board is less than (A). Then in STEP1 we erase all the earlier tagged a_i-squares and mark the a_1-square of the new trajectory [Fig. 31(c)]. We assign the length of the new shortest trajectory to the variable A. We iterate the process, not writing into STEP1 the a_1-square of any newly found trajectory having length greater than the current value of A, and replacing the current shortest trajectory by any new one of lesser length, i.e., writing in STEP1 the new a_1-square and purging the old values. As a result, a single inspection of the sheaf produces in STEP1 the a_1-squares of all

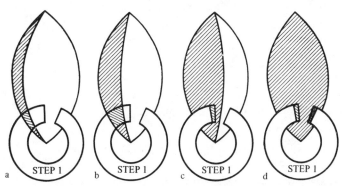

Figure 31   The analysis of trajectories in a sheaf by reference to their length.

shortest trajectories, which will all have the same length [Fig. 31(d)]. Among these we must find those that are forked.

2.8. Analysis of Trajectories for Forking

We now turn to the remaining sheaves of the given piece that are in the field. Of these we consider only those having an a_1-square that is already marked in the array STEP1 following the analysis of the basic sheaf chosen for inclusion in the search. Suppose that after that analysis we have written the marker 1 in the tagged squares of STEP1. If a trajectory in another sheaf has an a_1-square in a marked square, we increment the marker by unity. After we have inspected all the sheaves of the field belonging to the given piece, we find that STEP1 contains an array of integers, each denoting the number of sheaves having a trajectory that passes through the corresponding marked square, i.e., denoting the multiplicity of forking at the a_1-square of the basic sheaf. We now select the largest element or elements in STEP1. This integer is the coordinate of the most highly forked a_1-square of the shortest trajectories in the basic sheaf. Accordingly, the search will produce the move a_0-a_1. The procedure is shown in Fig. 32.

2.9. Retreat and Deblockade

Throughout this appendix we have used the term sheaf to mean an ensemble of those trajectories for a piece that have at least two squares in common—the initial and the terminal. In our model there exists yet another type of sheaf. We shall use the term sheaf of retreat trajectories (deblockade) to refer to the ensemble of all trajectories consisting of one displacement, having a common initial square a_0, and having as a final square any square on the board reachable by the given piece in one legal move. The criteria for inclusion of such sheaves in the search have not been considered

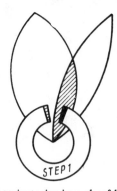

Figure 32   Analysis of trajectories in a sheaf by reference to forking.

in our current applications, nor has the formation of the corresponding fields.

The list of parameters of a retreat sheaf differs to some extent from the standard list. Figure 33 depicts the trace of a retreat sheaf; it is attached to the a_0-square of the sheaf. In the second slot we write the number 0 in place of the a_0-square, and in the third we write a 1. In the fourth we note the type of the sheaf, whether retreat or deblockade. The fifth slot is used primarily for the freezing–unfreezing procedures and for the calculation of T_x; in this slot we write the identification number (ID number) of the piece with whose trajectories the given sheaf was connected at the moment of its appearance, i.e., the ID number of the attacking piece or the piece to be deblockaded. We write a zero in the seventh and eighth slots of a retreat sheaf trace. In the seventh slot of a deblockade sheaf we write the coordinates of the a-square closest (in the trajectory) to the blockaded square of the a-square of the blocked trajectory—that is, the a-square on which the piece stood in the blockade.

For example, in Fig. 34 the trajectory Qc1-h6-h8 is blocked by the King at g5. The coordinates of the square c1 are written in the seventh slot of the deblockade sheaf for the square Z (g5). For a deblockade, the blockading piece must leave the trajectory, so that it is not newly blocked after the move. Therefore, in a deblockade sheaf, contrary to a retreat sheaf, some moves are eliminated from the consideration; e.g., in Fig. 34 we must suppress the trajectories Kg5-f4 and Kg4-h6 in the deblockade sheaf. To accomplish this, we write in the eighth slot the coordinates of the squares f4 and h6, the squares adjacent to the blocked a- (b-)square of the trajectory. The remaining slots are the same as in the general case.

Retreat and deblockade sheaves may also be stem sheaves of a field, and therefore all the field procedures considered in the first part of this Appendix are carried through in complete analogy for fields with stem trajectories

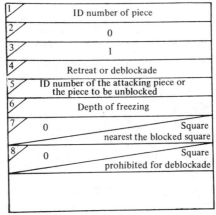

Figure 33   The trace of a retreat (deblockade) sheaf.

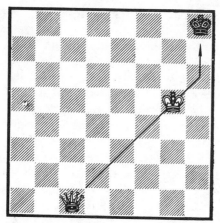

Figure 34   An example of a blocked trajectory.

belonging to retreat of deblockade sheaves. In executing a move on such a trajectory during the search we do not need to calculate the sheaf of the trajectory [3]; it suffices to use the total move generator (TMG) which we discussed in connection with the rapid determination of profitable captures (Section A1.2.6).

If we accomplish a deblockade before applying the TMG, the program often deposits pieces of the same color on the square written in the eighth slot of the trace and then removes them after applying the TMG. Thus the TMG treats the corresponding rays as established by its own pieces and the moves in these directions, as well as those forbidden by the rules of the game (in Fig. 34 these are the rays g5-h6 and g5-c1). Such a use of the TMG is due to A. I. Reznitsky. For retreat and deblockade sheaves, the degree of forking, and consequently the priority of inclusion of their trajectories, is calculated as described above.

As noted earlier, the necessity and feasibility of a deblockade is determined by analysis of its sheaves of trajectories for inclusion in the search. Every current shortest blocked trajectory is analyzed to determine the possibility of a deblockade. The total move generator is used in this analysis as well. If the deblockade is not possible, the blocked trajectory is no longer the current shortest one, and the procedure moves to the next trajectory of the sheaf. Otherwise we record the information needed for the deblockade (the parameters of the trace of the deblockade sheaf). Concatenating these traces in a chained list of cells, connected with the blocked square, we end the analysis of the trajectories in the sheaf if we find that the given trajectory is actually the shortest and is chosen as the move to be made (in the search).

The set of sheaves of trajectories, together with the retreat and deblockade sheaves, completely covers the set of lowest-level subsystems—the first level of the control system.

2.10. Pruning in the Presence of Branching in a Sheaf of Trajectories

The pruning which we considered earlier in this section is based on an analysis of the extent to which the goals of the second- and third-level systems are attained (i.e., in a field and in the MM). Here we consider a pruning based on an analysis of the extent to which the goal of the first-level system, the sheaf, is attained. Correspondingly, the result of this pruning influences only the search for a move in the given sheaf of trajectories. What we are dealing with here is stopping the inspection of moves in the trajectories of a given sheaf (at a given node in the search tree), and not the complete cessation of branching at the node followed by a backtracking ascent.

Suppose that we have arrived at some position during the search. The model selects a trajectory from some sheaf in order to make the succeeding move on it. Suppose further that we have already considered a variation, originating in the given position, in which a piece has moved along some trajectory in the selected sheaf. If the piece reached the terminal square (a_f-square) of the sheaf, it either left the trajectory (went into another field) or remained on the trajectory until it reached the end of the variation (i.e., was not blocked). Then movement along some other trajectory in the sheaf is senseless and the model prevents it. If, however, the piece was lost on this trajectory or was blocked, the model decides to move along another trajectory. Thus the suppression of branching in a sheaf depends on whether or not its a_f-square is attained, and if not, whether there remains hope of reaching it.

The criterion is simpler for deblockade sheaves. If in the given position there exists at least one deblockade trajectory and the blockading piece is not lost on the blockading square in the course of the variation, we inspect no other deblockades in the given position.

For retreat sheaves we always branch, since we must reach a safe square, and must test the possibility of continuing the attack.

PIONEER's pruning criterion is tested by analysis of the most recent of the tree branches constructed below the given node, and by study of the traces of the pieces involved, at the squares on which they stand in the given variation. The difficulty here is the following: at the moment of the analysis, the sheaf is among those included in the play in the field (here we are not considering the question of inclusion or exclusion), but it may turn out that it has only just now been included, and in the course of the given variation a piece may have moved along some trajectories, excluded from the field, that are forked with some that are included. In this case it is assumed that there has been no movement along the trajectories of the sheaf, and the decision is made to branch. To distinguish such situations, we inspect the twelfth slot of the parameter list of the sheaf. If during the search the sheaf is in an included field and the piece has made the move X-Y along some trajectory in the sheaf, the trace of the sheaf is associated with the square Y, and in its twelfth slot we write the address of the node in the tree corresponding to the position in which the move X-Y was made (see Section A1.2.11). If, however, the move was made via forking, and the trajectory itself belongs to an excluded field, we write a zero in slot 12.

The necessity of pruning branches in a sheaf became apparent in our experiments with PIONEER while it was solving chess problems.

2.11. The Structure of the Search Tree

Information on the search tree is used throughout the whole search and in the minimax process. As opposed to the majority of today's chess programs, PIONEER saves the entire tree in the computer memory, including branches corresponding to the current optimal variations. This turned out not to be difficult, since in none of our experiments did the tree contain more than 200 half-moves. The structure of the tree must meet the requirements of both descent and backtracking, i.e., establish in the backtracking the position at each node, the computer current material balance m_T, the minimax score of the variations, the erasure of certain branches during the backtracking in the pseudosearch, the liquidation of portions of the tree when a move has been made on the board, and "going along the trail".

All these requirements are met by a linked list. The nodes of the tree are numbered sequentially in the order of their formation during the search of variations. The serial number of the node yields the address of the corresponding information concerning the search tree [Fig. 35(a)]. It coincides with the column number in a two-dimensional array where the information about the node is stored. The structure of the standard column is shown in Fig. 35. In particular, the column contains the address of the immediate ancestor of the node and of the associated sibling node. The given node is obtained from the ancestor node after the move X-Y. If the move is

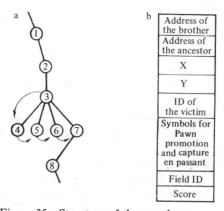

Figure 35   Structure of the search tree.

produced by a capture or by promotion of a Pawn, the information about the event is written in the corresponding column.

This tree structure allows us to determine, for any node and at any time during the search, what branches leaving the node have already been constructed, i.e., what moves have been made in the given position during the search. For this it suffices, given the number of a node, say 3 in Fig. 35(a), to increment it by 1 to obtain the address of its child. Here we find information about the move that led to this node (X-Y). We write the move in a special array. In the same column we find the address of the sibling [in Fig. 35(a) this is the node 5]. From the corresponding column we transfer information about the move to the special array, and so on. Thus we arrive at node 7, which has no siblings. In our special array we find a list of the moves that have already been made during the search at the given position.

In forming the tree we do not save positions corresponding to nodes that have been constructed. We save only the current position corresponding to the node being studied at the moment. Therefore we must record the position at each node, for use in backtracking during the search and in minimaxing. This is done by a procedure which changes the current position whenever a move is made, either forward or backward, in the search of a variation. The same procedure recalculates the current material balance m_T. While running, it uses information attached to the nodes of the search tree.

We have already noted that in the course of its work the program requires frequent reconstructions of the search tree. These consist of the removal of already constructed branches or whole subtrees. The tree structure we have adopted allows us to reconstruct the tree by renumbering the nodes. An example is shown in Fig. 36. Here we wish to remove the subtree consisting of the nodes 4, 5, 6, and 7. In essence, the program takes a census of certain columns contained in the two-dimensional array in another column.

2.12. A New Content of Known Procedures

The procedures described in the first part of this Appendix are new in comparison with the other known models of the game of chess. In the

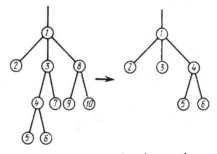

Figure 36   Restructuring the search tree.

second part we have collected descriptions of more or less standard operations, used in most chess programs, or even more widely in the solution of many enumerative search problems. By this we mean the search for moves, breaking off and scoring variations, minimaxing with pruning, and assigning the priority of moves in the search. But in our model, which considers a search as the hunt for the optimal strategy of a three-level control system, these operations acquire a wholly new content. This assertion is supported in particular by the dimensions of the search trees obtained in our experiments (of the order of 100 half-moves) and by the fact itself that we found solutions to problems in which the solution was 25 plies deep.

The application of the same formulae in the known chess programs leads to no such results. This is connected with the fact that none of these programs, as far as we know, model the game as a multi-level control system with an inexact goal, but rather as a one-level system provided with the rules of the game and with mate as its goal. Such a system is difficult to control because of the astronomical dimensions of the tree of variations in the search for an optimal strategy. The effect of applying the entire collection of the above-listed procedures within the limits of a one-level control system in the absence of an inexact goal scarcely justifies the effort invested in developing optimal search algorithms.

In the model we have been discussing, the role of the procedures described in Section A1.2 is not limited to shrinking the search tree. By influencing the search, they exhibit an influence on the interaction of the several levels of the control system: the formation of fields (level 1), the inclusion of fields in the search, and hence the provision of inverse feedback to the subsystems. In particular, the breaking off of variations (by the breakoff criteria) and their assessment by the scoring function are two of the most important aspects of the inverse feedback in a control system characterized by a search. The goal of the game answers the question—What is our aim?—and the scoring function tells us how successfully the goal has been realized.

APPENDIX 2
The Positional Estimate and Assignment of Priorities

M. A. Tsfasman and B. M. Stilman

The concept of the positional estimate plays an important role in chess programs. In algorithms involving an exhaustive search to a fixed depth, with the branch-and-bound method for pruning, the positional component of the scoring function is the most complicated of all components. The scoring function in such programs is a first-degree polynomial in several variables; its first term is proportional to the material balance in the current position, and the remaining terms depend on such factors as control of the center, pawn structure, etc.; taken together, they form the positional component. Perfection of the known programs amounts to perfection of the positional component of the scoring function by the inclusion of the new chess factors.

Such an approach to the positional estimate is static: factors that may be positive in most positions turn out to be negative in some cases, and the positional estimate is false in these cases. The positional estimate proposed by Botvinnik is based on the control of squares in trajectories. It is on the one hand common to all positions and on the other specific to each, since it is computed only for squares contained in the mathematical model (MM), which is in turn uniquely determined for each node in the tree. Thus, in the present model, the positional estimate is proportional to the ratio K_w/K_b, where K_w and K_b are the number of a-squares in the nonfrozen trajectories included in the play in the fields controlled by White and Black, respectively. We will show later how these a-squares are to be chosen.

The computation of the positional estimate is basically connected with two problems. We must first define the nonfrozen trajectories of the pieces and mark their a-squares. Then for each trajectory and for each a-square on it, we develop a list of the pieces lying one move away and compute the outcome of the optimal exchange on the given square. The square is traversable if the result of the exchange is advantageous to the side possessing the trajectory. If the given square is traversable and belongs, say, to White, K_w is incremented by 1. The treatment of Black is similar. The analysis of the a-squares of trajectories is carried on until the first non-traversable square is encountered.

Since PIONEER saves information about trajectories in packed form, or more precisely in the form of lists of sheaves of trajectories (see Appendix 1), we must unpack it in order to locate the a-squares; that is, we must extract the information about the sheaves from their traces and then apply a procedure for constructing trajectories, using this information as input parameters. This unravelling of all non-frozen trajectories in the included fields is a time-consuming process. One might hope to avoid it at some nodes of the search tree by noting that the position in the variation being searched is changing only sluggishly and that therefore the positional estimate need not be recomputed at every node, but only corrected from move to move.

But in fact the positional estimate exhibits great variability, since the MM does; the latter can change essentially, not only from one node to another, but even at a single node from moment to moment (for example, by the inclusion of new fields). This, in particular, emphasizes the uniqueness of our positional estimate for each position.

Accordingly, we cannot avoid unravelling the sheaves at every node of the search tree. It would be desirable to combine the calculation of the positional estimate with other operations that must be executed by the program at every node. We shall use the procedure for selecting a move in accordance with the assigned priorities. The procedure is as follows:

In the first version of PIONEER, the procedure consisted of repeated unravellings (according to the number of priorities) of sheaves while looking for a trajectory with given properties (see Appendix 1). The current version specifies a single unravelling of all sheaves, calculating the priority of each trajectory and then choosing the one with the highest priority. For a given piece, we define all sheaves in the fields that are included in the play. We unravel each sheaf in turn and determine the priority of each trajectory that has not yet been traversed (see below). An integer characterizing the priority is written into the special 8×8 array, in the a_1-square of the given trajectory. If we find that the a_1-square of a trajectory reached in the unravelling of a later sheaf coincides with a square already marked in the 8×8 array, we will have found a trajectory forked with one of a preceding sheaf. Therefore the priority is incremented, that is, we add 1 to the integer in the marked square. After we have unravelled all the sheaves for the given piece, the 8×8 array will contain positive integers for certain squares. We choose the square marked with the largest integer and store three quantities: X, the coordinate of the given piece; Y, the coordinate of the given square; and V, the integer written in that square.

We purge the 8×8 array by writing zeroes in all squares and repeat the operation for the next piece. If the number written in the selected square is larger than V, we store the new values of X, Y, V; if not, we pass to the next piece. As a result, after we have processed all the pieces for a given side, we have the three numbers X, Y, and V specifying the origin, destination, and priority, respectively, of the move having highest priority in the given position. This system allows us to compare priorities, not only for moves of a given piece, but also for moves of several pieces. We could not do this in the earlier versions of PIONEER. Moreover, while computing the priorities of the various trajectories, we can take account of diverse factors: the significance of the field (vulnerability and value of the target), the vulnerability of the trajectory itself, the length of the trajectory, and its forking index.

Now let us return to the computation of the positional estimate and show how the necessary procedure is included in the above process for selecting a move. After we have calculated the current trajectory with all its a- and b-squares on the real board, we must inspect its a-squares, beginning with the a_0-square, and apply at each of them the optimal exchange algorithm up to the first non-traversable a-square. It is easy to see that the optimum exchange procedure is the innermost cycle in the whole algorithm. This places heavy demands on it with respect to speed.

It is organized as follows for application to a given square in the trajectory of a given piece (pieces of the same color as the given piece will be called friendly, those of opposite color, enemy):

First we identify all pieces that can take part in the exchange on the given square, namely those that lie one move away. The trajectory of such a piece must be either free or blocked only by pieces that will take part in the exchange. Such blockades may be of the following types: Bishops and Queens may block one another, Rooks and Queens may block one another, and Pawns may block Bishops and Queens.

Therefore, by inspecting diagonals, verticals, horizontals, and possible posts for Knights, we can immediately identify all possible pieces that can take part. At the same time, we identify all blockading pieces. We of course break off the inspection of a diagonal, vertical, or horizontal whenever we find a piece that cannot take part.

Nowhere in this procedure do we need to compute a trajectory; this means a great saving in processing time.

We have now identified all pieces active in the exchange. We now use only the corresponding lists, and we may forget the real position on the board.

We are interested in the order in which the pieces will enter the exchange. From the viewpoint of the result of the exchange, the optimal order is that of increasing value, taking into account the fact that a piece can take part only after pieces blocking it have been removed.

Let us show that no other ordering will improve the result of the exchange. Choose another ordering, and suppose it is optimal and different from the ordering by increasing value. Choose two pieces of the same color that are neighbors in the list, and suppose the first has higher value than the second. We compare this variation with the one we obtain by inverting the order of usage of the two.

Let the first piece be captured. The enemy may break off the exchange; the result is the same as though we had taken a piece with lesser value. If the enemy continues the exchange, we may break it off; then we have lost the difference between the values of the interchanged pieces as compared to that obtained in the original order. If we continue the exchange we obtain the same result as we get in the original ordering, provided the enemy also continues, and we lose the difference in values if he breaks off.

This proves that the contemplated ordering does not improve the result of the original optimal variation.

Generally speaking, the blockade considered above may alter this principle. However, the customary scale of values of the pieces is such that the blockade of a trajectory belonging to a piece with lesser value than the blocking piece is possible only if the latter is a Queen; among the pieces active in the exchange, the value of a Queen plus the value of any other piece is not less than the sum of the values of two arbitrary pieces. This circumstance lets us extend our principle to the case of blocking pieces (the King cannot take part as an active blockader in an exchange).

We order the friendly pieces by increasing value, and do the same for the enemy pieces; then we compute the proper result of the exchange by a minimax procedure applied to the time when either side may break it off. We assume that the enemy is obliged to make at least one capture if he can do so. If he has no piece that can make the capture, the square is marked as traversable.

We take account of the following rules of chess in the minimax calculations: (1) the King cannot take part in an exchange unless he is the last to make a capture; (2) the square on which the exchange takes place may be one on which a Pawn would be promoted, in which case the value of the Pawn increases; (3) the first move in an exchange may be a capture en passant, which must also be taken into account.

Note that the result of an exchange is defined for a square in a trajectory rather than for the square on the board where it occurs.

This procedure for computing the result of an exchange on an a-square of a trajectory is used not only for the positional estimate, but also in determining the priority among trajectories. The priority is a linear function of a number of variables. One of them corresponds to the vulnerability of the trajectory; this means that all the squares in it are traversable, i.e., are controlled by the side to which the trajectory belongs. Clearly, with such an exchange procedure in the program, we can compute the positional estimate and at the same time determine the vulnerability of the trajectory for which we are analyzing the a-squares. This refers to attack trajectories; we are not concerned with the vulnerability of the victim's retreat squares.

In computing the priority of a given trajectory, it makes sense to calculate the vulnerability of the corresponding stem trajectory, aside from the vulnerability of the trajectory itself, which may, for example, be a denial trajectory in the field of the stem trajectory. Doing this at the same time as the priority of the given trajectory is being calculated is a complex and ineffective task, since it would be necessary to unravel the sheaves of another piece. It is much more convenient to determine the vulnerable fields by a separate special procedure. Therefore we first determine the priority of a field, and then the priority of the trajectories in it.

Before we determine the priority-driven choice of a move, we apply the procedure for defining the vulnerable fields. This unravels only the sheaves of the stem trajectories of the attack fields and computes their vulnerability, using the principle of optimal exchange on a square. The information on the priority of the stem trajectory of an attack field is written in the germ of the field, i.e., in the trace of its stem trajectory at the a_0-square (see Appendix 1). The same procedure determines the number of vulnerable fields and the total length of their stem trajectories; these quantities are necessary for the computation of the positional estimate. Once this is done, the procedure for choosing a move according to a priority determined for some trajectory finds no complexity in the resolution of the priority of the corresponding field. It is only necessary to refer to the germ of the field in question and extract from it the information on its priority.

The procedures described in this Appendix are implemented in PIONEER via two functions: computation of the positional estimate and determination of the priority of a move for inclusion in the search. Both are founded on the determination of the result of an exchange on a square. It is worth noting that both positional estimate and priority are computed without using precise numerical characteristics. The relationship between the positional estimate and material, and the interaction among the components of the positional estimate—vulnerability, length of trajectory, forking, etc.—are adopted temporarily as approximations, which will be refined in further experiments with PIONEER.

APPENDIX 3

The Endgame Library in PIONEER (Using Historical Experience by the Handbook Method and the Outreach Method)

A. D. Yudin

3.1. Introduction

As we have said earlier, the development of PIONEER posed the problem of making its content as close as possible to the content of the "program" of a chess master. The solution of this problem involved the development of an information system called "Historical Experience".

Any information system is worth developing only if it can be effectively used, and the solution of the problem presented by this fact is the most important factor in the development process. Only if there is reason to believe, and only if experiment bears out the expectation, that the problem can be solved is there a basis for going ahead with the development.

In our case we are concerned with the usefulness of developing the handbook-information system that we called "Experience of the Past" (EP) for a computerized chess program by modelling the thought process of a chess player. In the course of a game, a master constantly refers to his own experience and that of others. He does this at any node of the search tree where he considers it worthwhile. It follows that to increase the effectiveness with which experience can be used, it is necessary to have at most a small number of nodes in the search tree, and necessary for the program to use only small amounts of the computer's resources. Satisfaction of the second requirement is wholly in the hands of the developers of the EP system; the first requirement, however, is inseparably connected with the chess algorithm adopted. By using Botvinnik's algorithm, we obtain a humanly small search tree.

The chess player's program may be provisionally split into two parts: (1) the search for a move in the original situation and (2) libraries of openings, middle games, and endgames, plus a program for using them. The division is provisional because at times these two components work in parallel, in mutual cooperation. Handbook information is often used in the search for a move. In this Appendix we shall consider the development of the library of endgames used in PIONEER and the algorithms for the use of this library by the handbook and target-seeking methods. We also deal with the implementation of these algorithms in the program.

Since the pieces in PIONEER move in trajectories in accordance with an adopted goal, the problem of using an endgame library can be solved by modelling the behavior of a chess master. In the course of a game, a master not only looks for a coincidence between a position on the board (or in a search tree) and a library position, but also tries to make use of the latter if he finds it. (The search for library positions will be dealt with in Sections 3.10–3.17.) PIONEER does this also. After finding an advantageous and similar library position it looks for trajectories in which its pieces can move, and moves them so as to reach the target position. Once reached, this advantageous position coincides with the position in the search tree, the positional estimate is known, and the variation is broken off.

3.2. Formulation of the Problem

An endgame manual usually presents positions and corresponding varia-tions. A strong master, however, does not remember these variations. He remembers the base positions, the nodal positions, the scores, and, if necessary, any difficult initial moves. He leaves the rest to his algorithm for selecting a move.

It is therefore useful to omit the variations from the endgame library, keeping only the nodal positions, their scores, and initial moves if these are difficult. This makes for an essential simplification in the methods for developing a library and decreases the volume of stored information [4].

Thus the master (or program) has the following task: He has a concrete endgame position, arising either in the play or in some variation during the search. With the help of the library he must score it (win, draw, lose) as the move shifts from one side to the other and then, if necessary, choose the best first move.

3.3. Configurations

By the configuration of a position involving N pieces we shall mean a set $D_1, D_2, \ldots, D_{N-1}$ of coordinate differences $L_i - L_{i-1}$ for $i = 1, 2, \ldots, N-1$, where L_i is the linear coordinate of the ith piece in the position (it may range from 1 to 64).

Let us imagine a position in a technical endgame. Further, having fixed the relative locations of the pieces in the position (the configuration), let us move the position vertically and horizontally to all possible positions consistent with staying on the board and not putting pawns in illegal places. We now have a set of positions belonging to the given configuration. In our program development the task is to see how to describe this set of positions (there may be as many as 40) in the endgame library (together with their solutions: scores, first moves) in a form suitable for programming.

3.4. The Boundary Effect. Decomposition Formulae

The needed information an be written in compact form because of a phenomenon called the boundary effect. It was found that the set of positions is easily subdivided into four subsets that, from the viewpoint of their scores and decisive first moves, do not intersect:

(1) positions influenced by one of the vertical sides of the board;
(2) positions influenced by one of the horizontal sides;
(3) positions influenced jointly by a horizontal and a vertical side (the so-called corner positions);
(4) all other positions in the set not subject to the boundary effect.

The scores and first moves for a given configuration are constant within any one of the above subsets. Moreover, since the whole set is characterized by a single configuration, and the positional estimate has only three values (win, draw, lose), it follows that in the majority of configurations some of the subsets may be merged because of the identity of scores and first moves. Also, we often encounter empty subsets, containing no position whatever.

Thus, for the recognition of an arbitrary position in the set it suffices to store in the library a single exemplar, for instance from the corner subset (often, but by no means always, this subset degenerates into a single position), together with the set-decomposition formulae which characterize the limits of variation in the score. We shall give a detailed solution of the problem for one set by means of an example.

3.5. Symmetries

The use of symmetries is extremely important for the program:

(1) Flank symmetry—the reflection of a position in a vertical axis of the board;

(2) Color symmetry—reflection of the position in a horizontal axis, with corresponding interchange of colors;

(3) Diagonal symmetry—reflection of the position in the diagonals a1-h8 or h1-a8, for Pawnless positions.

Every square on the board is characterized by its two-dimensional coordinates x, y or by the linear coordinate L; the two systems are connected by symmetrization. Let x, y, L be the coordinates of a square containing a piece before symmetrization. Then for flank symmetry

$$L_f = L + 9 - 2x;$$

for color symmetry

$$L_c = L + 8(9 - 2y);$$

and for diagonal symmetry

$$L_{a1\text{-}h8} = 8L - 63y + 56 \qquad \text{and} \qquad L_{h1\text{-}a8} = 63y + 9 - 8L.$$

Note that for an arbitrary square

$$L_{a1\text{-}h8} + L_{h1\text{-}a8} = 65.$$

These relationships and a few others are used to form symmetrized configurations.

These symmetries are used either separately or together, according to the circumstances in which they are applied. Thus, the position White: Ke3, Rh1; Black: Kg5, Bh4, entered in the book in either direct or indirect form (by the use of formulae employing the boundary effect), characterizes 15 positions connected with the joint use of the symmetries.

3.6. The Structure of the Library. Classes. Coding of the Information

The structure of the library is as follows: All positions in a technical endgame are broken down into 31 classes with respect to material. For instance, "King and Pawns vs. King"; "King, Knight, and Pawns vs. King and Bishop", etc.

Thus, within each class the material balance is constant over all positions.

A class as stored in the library is represented by a rectangular matrix of dimension $m \times n$, where m is the number of positions in the class and $n = \text{Ent}[(N + 1)/2]$; N is the number of pieces in each position of the given class.

Each row corresponds to a single position. The information in the first position is coded as follows:

White: Kf2, Pg2; Black: Kc6, Ph4.

The solution of this position is as follows: White wins with 1. Kf2-g1; if Black is to move, it is a draw: 1. ...h4-h3. Let this position have index number *i* in the class "King with Pawns against King with Pawns".

The score *r* may take on the following values:

$$R = \begin{cases} 0 - \text{won}, \\ 1 - \text{drawn}, \\ 2 - \text{lost}. \end{cases}$$

The first $n-2$ positions (in the row of the matrix, not on the board) form pairwise united coordinates of pieces. The next two columns contain five-digit numbers (in the general case) representing moves preceded by the corresponding scores. Thus, the elements of the *i*th row in our illustrative matrix are, respectively: $M_{i1} = 1415$; $M_{i2} = 4332$; $M_{i3} = 1407$; $M_{i4} = 13224$.

In classes with *N* odd, the column $n-2$ contains the two-digit coordinates of the *N*th piece in the position.

To each position in a class there correspond formulae for decomposing the set of which the given position is the configuration symbol, the first moves, and formulae for correcting the score (see the example below).

With this structure, the insertion (deletion) of positions in the library amounts to the inclusion (removal) of a row or rows with respect to the class matrix.

Let us consider a search by the handbook method, and the operation of the boundary effect, as exemplified by a simple configuration (see the position depicted in Fig. 37 and the search method shown in Fig. 38). The positions in the table (except for the last column, which is introduced for perspicuity) are written in the library and correspond to positions in the diagram of the class "King and two Pawns vs. King and Pawn".

Figure 37   A sample position.

Using symmetry by flank and color, each position in the set is converted into four positions.

3.7. Organization of the Information in the Form of Two-Dimensional Tables with Subordination of Entries

The presence of the boundary effect allows us to put all the endgame handbook information into a number of two-dimensional tables with subordinated entries.

A two-dimensional table with subordinated entries is one containing two entries, one independent and the other dependent on the first. (See Table 1.)

In our case a single two-dimensional table corresponds to every class with the characteristic material balance. Figure 38 depicts the structural scheme of the program for searching such a table. The first required entry into the table is by configuration, that is, a search for coincidence between the coordinates of the position on the board (or in the search tree) and the relative coordinates of one of the library position-configuration symbols. If and only if a coincidence is found, control passes to the procedure for making the second entry into the table, i.e., finding a coincidence in absolute coordinates. Note that this part of the table (the greater part) does not exist in absolute form, but rather as a set of decomposition formulae, which are applied only after the coincidence of relative coordinates has been established [6].

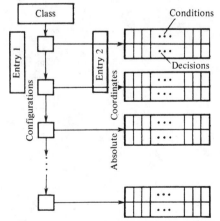

Figure 38 Structure of the search for data in a two-dimensional table with subordinated entries.

Table 1 Characteristics of a set of positions (for the position shown in Fig. 37).

Subset Number	Method for Getting Subset from Embedded Library Position	Positional Estimate		First Move		Number of Positions in Subset
		White to Move	Black to Move	White	Black	
1	Move down 1–3 squares	draw		&	&	3
2	Subset is empty					0
3	Same as embedded position	draw	win for Black	&	&	1
4	Move left 1–6 squares and down 1–3 squares	win for White		$l_0^{WK}-l_1^{WK}$, & $l^{WK}=l_0^{WK}+9$	&	18

Note: "&" signifies that the move is forced. There is no type 2 subset in the given example. The notation l_0^{WK} refers to the position of the White King before the first move; l_1^{WK} refers to the position after it.

3.8. The Algorithm for Using the Endgame Library (The Search for Exact Coincidence)

The complex of procedures that implement the algorithm is brought into action whenever a material balance corresponding (to within an interchange of color) to a class entry in the library is encountered in a board position or in a variation during a tree search.

Let us inspect the flowchart of the algorithm for using the endgame library (Fig. 39). The procedures for diagonal symmetry are omitted for the sake of simplification.

Procedure A computes the material balance, which is needed to determine whether color symmetry exists. All the later procedures assume superiority or equality of White; the library is constructed on this principle also. Therefore if Black is superior in the board or search position, a color symmetry is performed by procedure E, and a solution is found for the position with the colors reversed. Then procedure U inverts the transformation, the necessity of which is indicated by the value SC $= 1$ in procedure D. If the board or search position is in equal material balance, the library position may be sought for either the initial or symmetrized position. (SC $= 2$ in procedure C.)

Procedure F brings the initial position into a form suitable for comparison with the library entries, i.e., rows of the two-dimensional class matrix. Here the pieces are sorted and the initial position is coded. Then the class number is computed by procedure G, and procedure I calls the corresponding class into operating memory.

Let us look at procedure G. For every position in a class there exists some general characteristic, called the template, which defines the concrete relationship of the pieces.

The template for position A is an array U with 12 elements: $U(1) = 1$, $U(2)$ is the number of White Queens in position A; ... $U(12)$ is the number of Black Pawns.

The first six elements of the array $U_w(6)$ are called the White template; the second six, $U_b(6)$, the Black. These templates are stored for each class and represent its material characteristic.

Just such a template is prepared for the board or search position. The search for a class number by matching the templates is depicted in Fig. 40. This search may employ color symmetrization, depending on the value of SC.

The coincidence of the template of the initial position with one of the class templates yields the number of the class that procedure I calls into operational memory.

Procedure H employs symmetries to produce all possible positions from the initial position.

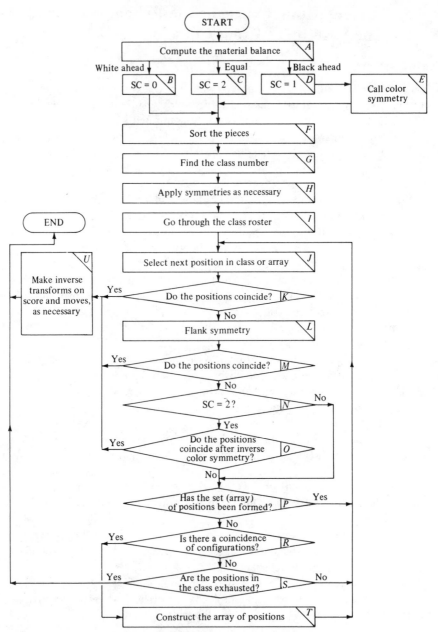

Figure 39   Flowchart of the algorithm for using the endgame library (the search for exact coincidence).

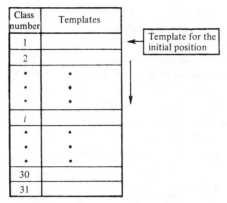

Figure 40   Schematic diagram for the search for a class number by comparison of templates.

When control first passes to procedure J, procedures K, L, M, N, and O carry out a search that is exact to within a color symmetry for a match between the initial position and the position-symbol of a set in the class (this is a search for the initial entry point into the two-dimensional table); when the search is successfully completed, control passes to procedure U, which reverses the score and first move if necessary.

Let us assume (what is in fact the most probable case) that no exact match (even taking account of symmetry) is found. Then the test P, which first takes control, yields a negative answer and passes control to the test R, which answers the question of the necessity for restoring or establishing a set with respect to the position symbols, i.e., with respect to the library positions that we have just now dealt with.

In order that a solution exist within the limits of the set for the given position-class, it is necessary and sufficient that the configuration of the position coincide with at least one of the symmetrized positions on the board or in the search. The proof of necessity and sufficiency follows in an elementary way from the definition of a configuration, the rules for symmetrization, and the process for forming the set of positions for a given configuration.

Thus, given a positive outcome of the test R, we have a full guarantee that the array or set of positions will not have been formed in vain (procedure T). This means that after all the positions in the set formed by procedure T are tested, control necessarily passes to procedure U because of the positive outcome of one of the tests K, M, or O. This constitutes the search for the second entry in the corresponding two-dimensional table.

Note that with the aid of the formulae for decomposition and for correcting the score and the moves, procedure T produces a set of positions with their solutions, which are inscribed in the two final columns of the matrix. After the necessary inverse transformations have been applied to the score and the move by procedure U, the problem is solved.

Now assume that the test R yields a negative result. Then control passes to procedure J; the next position-symbol is chosen, and so on.

If we have examined all the position-symbols in the given class and the test R has not yielded a single positive outcome, the test S gives a positive result, which means that no exact match exists; accordingly, the desired position is not to be found in the library. This happens often, since the library contains relatively few positions and the number of possible configurations is very large in a technical endgame, even though the number of pieces may be small. In this case we have recourse to the outreach method as applied to library positions (see Section A3.10 ff.).

The methods we have just described generate some 14000 positions (without taking symmetries into account) from a library of 633 positions.

3.9. Examples of the Operation of the Subroutine for Using the Endgame Library

Our experiments with the technical endgame library had two goals: (1) to test the efficacy of the subroutine for using the library when called from the subroutine for searching for moves; (2) to make an analytical test of the scores corresponding to the embedded information. We also tested the influence of the boundary effect on the score for the position, together with the use of the symmetrization mechanism.

After we had finished the chess portion of the work, i.e., after we built the library itself, we wrote a special subroutine to compute the solution of all the positions in the technical endgame library (with the aid of the decomposition formulae—see section 3.4) and printed out all the arrays and the solutions of each position in the library. One such array is shown in Fig. 41.

The embedded configuration is shown in the computer display. The first column of the table contains the serial number; the second contains the position of the White King; the third contains the score for White's move ("$+-$", win/lose for White; "$=$", a draw); and the fourth contains White's move. The name of the piece making the move is omitted; the code B9-B9 means that the indicated move that solves the position is not obligatory. The fifth column contains the score if the move is Black's; and the sixth contains Black's move.

These arrays were subjected to examination by tens of strong chessplayers; as a result, we eliminated a number of errors, unavoidably present in the construction of such a large library and of the corresponding subroutines.

Figures 42(a)–42(c) show the output from the subroutine for using the library. This is the solution of the symmetrized positions obtained from a single configuration. The interpretation of the captions printed above the

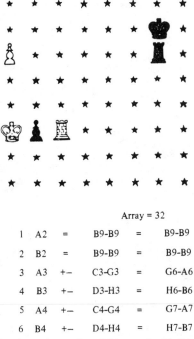

Array = 32

1	A2	=	B9-B9	=	B9-B9
2	B2	=	B9-B9	=	B9-B9
3	A3	+−	C3-G3	=	G6-A6
4	B3	+−	D3-H3	=	H6-B6
5	A4	+−	C4-G4	=	G7-A7
6	B4	+−	D4-H4	=	H7-B7

Figure 41   An example of an array of positions embedded in the library with their solutions.

diagrams is as follows:

New Position: A new position generated by the subroutine for analysis; a "−" sign before the piece code indicates a Black piece. The uppermost diagrams display the starting positions; those in the second row display the outcome and the position after the first move when White is to play; those in the third display the corresponding information when Black is to play.

3.10. Outreach for a Library Position

All the foregoing considerations relate only to exact matches between positions on the board or in the search and positions in the library. Let us now suppose that a position arising in a game or in some variation in the search tree is such that no matching library position can be found. How does a chess master act when faced with such a problem, and therefore how should a program modelling his thought proceed?

Capablanca showed [9] that a chess master not only goes over library positions in his calculations in order to break off a variation, but also tries

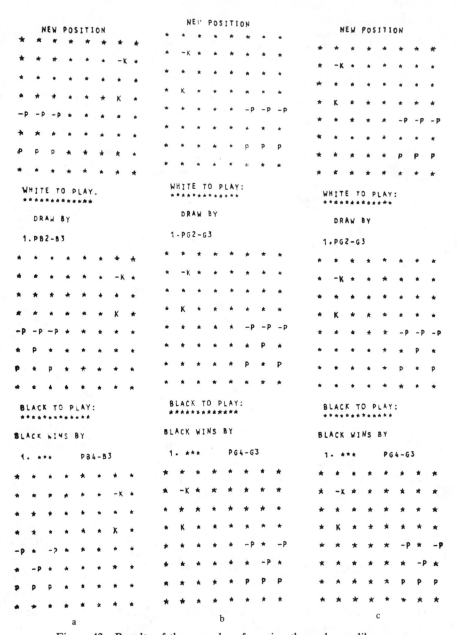

Figure 42   Results of the procedure for using the endgame library.

to use one or other of the library positions to his own advantage. The program should do the same.

Suppose that the configuration in the starting position has not been found among the position-symbols in the sets of library classes. Then a position, or group of positions, in the library that are very near the given position (on the board or in the search) should be found, and an effort to reach such a position should be organized, in the following way: When the neighboring position has a favorable score for our side, we include in the mathematical model some so-called planned trajectories for the pieces belonging to the active (target-seeking) side, i.e., trajectories leading from the starting position to the neighboring position. The active side forms its fields. Here the algorithm for using the library cooperates with the algorithm for searching in the starting position. In order to organize the search, we must include in our mathematical model the trajectories of all the pieces of the passive side that do not coincide with the library position.

It is important to note that these trajectories must be "anti-forked" in the sense that the passive side must move along them only in extreme cases. Moreover, in contrast to the fundamental principles of the move-search algorithm in the starting situation, no fields are formed to obstruct the movement of the passive side's pieces in the designated trajectories.

3.11. The Search for Nearby Positions

First we define the following notions:

Two positions A and B will be said to be *equal in material* if $U_i^A = U_i^B$ for all $i = 1, 2, \ldots, 12$, where U_i^A and U_i^B are the elements of the templates of positions A and B.

The non-coincidence s_{AB} of two positions A and B that are equal in material is defined as

$$s_{AB} = \tfrac{1}{2} \sum_{i=1}^{64} c_i,$$

where $c_i = 1$ if $a_i \neq b_i$, 0 if $a_i = b_i$.

Here a_i and b_i are the programmed codes for the pieces located on the ith square in the positions A and B, i.e., $a_i =$

11. if the ith square in position A contains a White Pawn,
13. if the ith square in position A contains a White Knight,
14. if the ith square in position A contains a White King,
15. if the ith square in position A contains a White Rook,
16. if the ith square in position A contains a White Bishop,
17. if the ith square in position A contains a White Queen,

22. if the ith square in position A contains a Black Pawn,
23. if the ith square in position A contains a Black Knight,
24. if the ith square in position A contains a Black King,
25. if the ith square in position A contains a Black Rook,
26. if the ith square in position A contains a Black Bishop
27. if the ith square in position A contains a Black Queen,
 0. if the ith square in position A is empty,

where i is the linear coordinate.

Let $T_1, T_2, \ldots, T_{s_{AB}}$ be a sequence of trajectories leading from position A to position B (i.e., trajectories in which the moves begin at A and end at B).

The difference D_{AB} between positions A and B of equal material is defined as

$$D_{AB} = \sum_{j=1}^{s_{AB}} l_j,$$

where l_j is the length of the trajectory T in half-moves for any $j = 1, 2, \ldots, s_{AB}$.

Two positions A and B of equal material are said to be neighboring if $s_{AB} \leq s_{max}$; $D_{AB} \leq D_{max}$, where s_{max} is the maximum admissible noncoincidence (number of non-coinciding pieces) and D_{max} is the maximum admissible difference in the positions.

As a first approximation it was assumed that the search for neighboring positions should confine itself to positions of equal material. This means that the class of position-symbol configurations in the library is uniquely defined as the class for which the material (taking account of color symmetry) coincides precisely with the material in the starting position (see Fig. 40). The values of the quantities s_{max} and D_{max} serve as running limits on the search.

Then the problem is as follows: We have a starting position A and wish to find a group of library positions $B_1, B_2, \ldots, B_n$, each near A in the sense defined above.

Note that here we are not analyzing the relationship between the score of A and the scores of the $B_1, B_2, \ldots, B_k$; that analysis will be carried through later. We do not assume that we shall subsequently organize an effort to reach all of the B_i.

We call the positions B_i *candidate targets*. Obviously, at the outset all positions in all sets defined by the position-symbols of the given class are candidate targets.

One of the tasks facing the search for neighboring positions is to make the maximum reduction in the number of candidate targets without using an awe-inspiring machinery for finding the trajectories. We have found a number of constraints, which we call filters, that allow us to reject a large number of the sets of library positions by finding positions in the sets that, although near the starting position, can be rejected at the configuration level. These filters are of the following types.

3.12. Filter with Respect to the Pawn Structure

This filter operates only when the starting position (and therefore the library position also) contains two or more Pawns of either color.

We have the starting position A and a position C which typifies the set of positions belonging to the given configuration. We form the Pawn template for position A as follows: We project the entire Pawn structure on the first rank and assign to the eight squares the values

$$SH_i^P = 0 \text{ (if there are no Pawns in the } i\text{th file)}$$

$$= 10p_i^w + p_i^b \text{ (if the } i\text{th file contains a Pawn or Pawns)},$$

where p_i^w is the number of White Pawns in the ith file and p_i^b is the corresponding number of Black Pawns; $i = 1, 2, \ldots, 8$.

This process yields an eight-element array SH^P. We further select the non-zero elements by eliminating all zeroes from the left-hand side until we reach a non-zero element $SH_i^P \neq 0$, and from the right-hand side so that the final element is $SH_j^P \neq 0$ ($j \geq i$). The array SH^A so obtained is called the Pawn template for position A, and is assigned the length $l_A = j - i + 1$.

The Pawn template SH^C, of length l_C, is constructed in the same way for position C.

We shall prove the following assertion. There exists no sequence of trajectories leading from a position A to an arbitrary position in the set characterized by C if: (a) $l_A \neq l_C$; or (b) $l_A = l_C$, but there exists at least one value of i in the set $i = 1, 2, l_A$ such that $SH_i^A \neq SH_i^C$.

PROOF. Let C_1 be one of the positions in the set characterized by the position-symbol C. Since the positions A and C are equal in material, so are A and C_1.

Suppose that one of the conditions (a) or (b) holds.

We assume that our assertion is false, i.e., that there exists a sequence of trajectories carrying A into C_1. Then if either condition holds, at least one trajectory must change the number of Pawns in some file by at least one unit. This follows from the fact that if one of the conditions holds, the Pawn templates do not coincide. But a Pawn can move from one file to another only by making a capture or captures. This contradicts the fact that A and C_1 are equal in material, and this result holds for any C_1 in C. We have arrived at a contradiction, and therefore the assertion is valid.

If the templates SH^A and SH^C do not completely coincide, all the positions in the set characterized by C may be excluded from the list of candidate targets.

It is worth noting that all the arguments we have used are valid to within a symmetry. This means that if Condition (b) holds, we must form the flank-symmetrized Pawn template SH_{fl}^C by the formula $SH_{i(\text{fl})}^C = SH_{l_C-i+1}^C$ ($i=1,2,\ldots,l_C$) and test the condition (b) for SH^A and SH_{fl}^C.

If Condition (b) is not satisfied, i.e. $SH^A = SH_{\text{fl}}^C$ for all $i=1,2,\ldots,l_A$, the configuration C is listed among the candidate targets, and the necessity of performing a flank symmetry is recorded.

Furthermore, if the White and Black material are identical in the given class, in the chess sense, we must take account of the possibility of a color symmetry for SH_c by means of the formula

$$SH_{i(\text{co})}^C = \begin{cases} 0 & \text{if } SH_i^C = 0; \\ 10p_i^b + p_i^w & \text{if } SH_i^C = 10p_i^w + p_i^b. \end{cases}$$

The relative Pawn structure filter yields a great reduction in the number of candidate targets.

3.13. The "One Color–Different Color" Filter

This filter acts only when each side has a Bishop and they are on diagonals of opposite colors.

Suppose given an initial position A, and denote the two-dimensional coordinates of the White Bishop by x_w, y_w and of the Black by x_b, y_b.

It is clear that if x, y are the two-dimensional coordinates of a square on the chessboard, the sum $x + y$ is even for any White square and odd for the Black squares. Therefore the sum $Z_A = x_w + y_w + x_b + y_b$ is even if the opposing Bishops occupy squares of the same color and odd for squares of opposite color.

Clearly no trajectories can lead from the position A to a position C if Z_A and Z_C are of opposite parity, i.e. if $Z = Z_A + Z_C$ is odd. It is worth noting that this filter does not require symmetrization, since no symmetrization performed on C can change the parity of Z_C nor, therefore, the parity of Z.

Thus, after we have initially filtered the list of configurations, we have a list of candidates $C_1, C_2, \ldots, C_m$ such that candidate targets can be found only in the sets of positions characterized by these configurations.

3.14. Filters Within a Set of Positions

We may now proceed directly to the formation of a list of candidate target positions.

We have an initial position A and a sequence of candidate target configurations C_i. All the positions in the set characterized by C_i are candidate target positions. The following constraints (filters) help in reducing the number of the latter.

The symmetrization algorithm is essential in all phases of the search for neighboring position. Throughout the remainder of this discussion, however, we shall not mention it explicitly.

From what we have said in Section 3.12, it follows that the sequential examination of the list of candidate targets results in the selection of only those positions in the set for which the files of all the Pawns correspond exactly with the files of the Pawns in the initial position A. This is not all, however. We also require that the coordinates y_j^A of the White Pawns in position A satisfy the conditions $y_j^A \leq y_j^C$, and for the Black Pawns the conditions $y_j^A \geq y_j^C$. Here the y_j^C are the second coordinates (vertical) of the corresponding Pawns in the selected position taken from C_i. In other words, no Pawn in the initial position may be further advanced than the corresponding Pawn in the candidate position, else a trajectory would be required in which a Pawn moved backward, violating the rules of chess.

Again, if there is a Bishop in the candidate position, it must be on a square of the same color as in position A.

After all the positions in the set have been filtered, we have the initial list of candidate target positions.

We must note in the conclusion that the program implementing the algorithm that searches for neighboring positions does not record the current candidate target positions, but rather records only the locations, in the program of the decomposition formulae that generate these positions by use of the symbols of the corresponding sets.

3.15. Finding a Group of Neighboring Positions

When the list of candidate target positions is small, we can make a direct test of the nearness of each to the initial position.

First we test to see whether the discrepancy exceeds the maximum, s_{max}. The list is again greatly reduced. Next, when the number of non-coincident pieces does not exceed the established limit, we resort to the algorithm for generating trajectories. After having found trajectories for the non-coinciding pieces, i.e. those trajectories that carry the neighboring position into the initial position, we test to see whether we have exceeded the maximum allowed distance between the positions D_{max}. Then we may decide whether to include the candidates in our final group of neighboring positions, which we denote by $B_1, B_2, \ldots, B_n$.

Note that the quantities s_{max} and D_{max} may vary, depending on the computer's resources of time and memory. In our opinion, a chess master varies them in the same way. Under time-pressure or memory-overload even a world champion does not allow himself large values of s_{max} or D_{max}.

The flowchart of the algorithm for near-neighbor search is shown in Fig. 43.

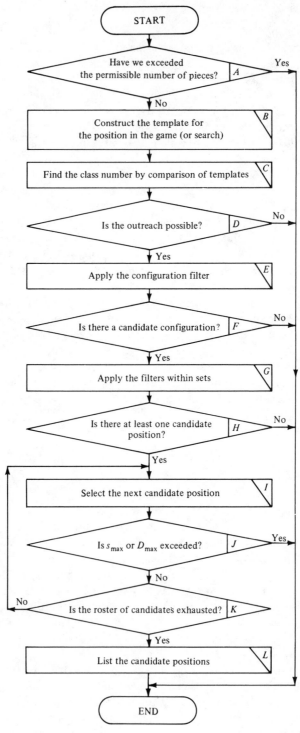

Figure 43　Flowchart of the algorithm for finding the neighboring positions in the outreach method.

3.16. Implementation of the Outreach Method. Anti-outreach

Let us again recall what is meant by reaching out from an initial position A to a neighboring library position B. To organize such an outreach means to assign priorities to trajectories that lead from A to B and include them in the mathematical model on the basis of information in the technical endgame library about the score and the decisive moves for the side that has the move in position B.

The implementation of the method based on outreach toward a library position may lead to a directed formation of the mathematical model and the search tree. But as soon as coincidence is established, the handbook method takes over, based on complete identity of positions; the score is known, and the variation is broken off.

Let us now introduce the notion of an anti-outreach.

Suppose that in the initial position A there is a win for White at some White node in the search tree (i.e. the move belongs to White), and that for some neighboring position B the outcome is a draw. Suppose also that for some reason or other the mathematical model includes trajectories leading from A to B. Then anti-outreach means the forced lowering of the priorities for moves in those trajectories. In this sense, anti-outreach enables us to avoid traps and snares.

3.17. Entry from the Move-Search Routine

Let us now see how the outreach subroutine is entered from the move-search routine in the original situation. If we were to test for coincidence of the current position and a library position at every node of the search tree, much time would be consumed. (Such a procedure is characteristic of the handbook method.) But if we test for neighboring positions at every node we are infringing on the algorithm for the inclusion of fields. Before going to an outreach we must determine what additional benefit we would derive, since every outreach is connected with the inclusion of fields and an enlargement of the search tree. Therefore we perform our outreaches in the same sequence as the inclusion of fields, i.e. only during backtracking up the tree.

We recall that if the mathematical model of the position contains not even one planned trajectory of the passive side, the outreach to the position is not to be organized, and the position is removed from the list of candidates.

3.18. Outreach and Exact Coincidence

Let us list some observations on the similarities and differences of the handbook and outreach methods.

From what we have said above it is clear that the search for exact coincidence of two positions is a particular case of the search for neighboring positions. In fact a library position coinciding exactly with the initial position may be described in general as a neighboring position.

Suppose given an initial position A and a library position B equal in material to A. Suppose further that $s_{AB} = 0$ and $D_{AB} = 0$. Then, firstly, B coincides exactly with A, and secondly, by the definition of a neighboring position, it is a neighbor of A. Therefore a zero discrepancy and distance between two neighbors implies their exact coincidence.

In this respect the handbook and outreach methods are alike.

The difference between these two methods for using historical experience lies in the fact that the handbook method, as a particular case of the outreach method, does not require the use of the search algorithm in the original situation; more precisely, it does not require the mechanism for generating trajectories nor does it extend the search tree for variations. For these reasons the handbook method is simpler, in that it requires less expenditure of the resources of the computer. The same reasons also account for the order in which we have presented the use of these methods for building the technical endgame library and the algorithms for exploiting it.

3.19. The Potential for Wide Use of Library Rules in PIONEER

In conclusion, we shall deal with a portion of the endgame library that is not used for complete or incomplete comparison of positions encountered in the search for variations with those met earlier. This is the so-called library of rules. In the course of our development of decision theory as applied to an arbitrary domain, we elaborated a set of methods usable for scoring a situation while it is under construction, without further formation of the tree of possibilities; instead we use certain criteria, or rules, having the character of strictly proven theorems. In chess, there are no exceptions to these rules. The more such rules a chess master knows, the more likely he is to break off a variation in the search tree rapidly by using them.

The development of any program based on the construction of a tree of variations and the use of the minimax procedure on it inevitably raises the problem of forcing the breakoff of a variation before reaching its limiting depth, by the use of the various criteria. The first solution to this problem was developed in PIONEER. The question concerns a breakoff in accordance with certain library rules, rather than a breakoff connected with the goal of the play (by the search routines in the original situation). It is obvious that the greater the number of variations we can break off in this way, the smaller the final size of the search tree. Therefore we quite naturally wish, first to elaborate and formalize a large number of such criteria for breaking off a variation, and then to use them at as many nodes of the search tree as possible.

On the other hand, the time required for solving the problem may grow very rapidly as the number of nodes at which we use the criteria increases. The effectiveness of the method is closely dependent on the size of the search tree itself. Since the test for these criteria at a given node consumes some time, even though small, it is clearly senseless to make the test at all nodes of a large tree. We are speaking here, not of the formal pruning by the α-β-procedure, but of the use of breakoff criteria connected with a specific task, in our case with the use of a library of chess rules.

Clearly, the use of a library of rules is of dubious effectiveness for the majority of the existing chess programs, which are dedicated to the principle of a full-width search and consequent production of enormous search trees.

The picture is quite different in the development of PIONEER. Here we have to do with a search tree that is small and, what is extremely important, narrow—containing some 100 nodes. With the algorithm we have adopted, the widespread use of library rules is quite possible. Note that there is an obvious inverse feedback at work here: an effective use of the rules (methods of play at various stages of the game) allows us to shrink the search tree even more stringently, which not only speeds the solution of the problem but qualitatively influences the effectiveness of the solution.

Thus, in the solution of inexact search problems by means of Botvinnik's algorithm there is an interaction between the small size of the search tree and the use of historical experience and accumulated knowledge of the substance of the investigation, in this case the game of chess.

3.20. The Breakoff Criterion Based on the Rule of the Square

During its many centuries of development the theory of chess has produced an enormous number of methods of play, of all possible types, allowing the calculation of variations to be shortened.

In the development of PIONEER we found it expedient to embody a large number of these methods, formalized as criteria for breaking off a variation, in order to increase the effectiveness of the solutions to our problem.

As an example, let us consider the formalization and extension of the so-called "rule of the square" in a Pawn ending. We omit a detailed explanation of its essence, since it is discussed in detail in many special texts; it has to do with whether a King can prevent an opposing Pawn from reaching the eighth rank successfully.

Since PIONEER can play either side, we shall not speak explicitly of either White or Black, but rather of the $(+)$ and $(-)$ sides. We adopt the following notation:

K($+$)—King on the $(+)$ side;
P($+$)—Pawn on the $(+)$ side;
K($-$)—King on the $(-)$ side;
P($-$)—Pawn on the $(-)$ side.

Absolute Criteria. Material K($+$), P($+$); K($-$)

1. If the K($-$) is not within the square of the Pawn P($+$), the side $(+)$ wins.*
2. If the K($-$) is in the square of P($+$) and nearer to P($+$) than is the K($+$), the game is drawn.

Here, and later, we shall mean by "nearer" that

$$|K(-)-P(+)| < |K(+)-P(+)|,$$

where the terms on either side denote distances in half-moves between the corresponding pieces. These distances are computed in an elementary way from the coordinates of the pieces. The formula is independent of the assignment of the move.

The First Sufficiency Criterion. Material: K($+$), P($+$); K($-$), P($-$)

Suppose both Kings are within the squares of the opposing Pawns. In order that one side may be sure of a draw it is sufficient that his King is closer to the enemy Pawn than the enemy King is to the same Pawn. For instance, if

$$|K(+)-P(-)| < |K(-)-P(-)|,$$

after taking account of the assignment of the move, the $(+)$ side is sure of a draw.

Let us prove this assertion. Suppose the given condition holds. If the $(+)$ side moves only his King, and moves it in the direction of the enemy Pawn in such a way that the distance decreases with each move (which he can do because he is within the square of the enemy Pawn), the K($+$) can capture the enemy Pawn without hindrance before it is promoted. This guarantees a draw.

---

*Here, and in what follows, in defining the location of the King in the square of the Pawn, we must take into account the possession of the move. For instance, depending on which side has the move, the King if on the boundary may be either in or out of the square.

Second Sufficiency Criterion. Material: $K(+)$, $P(+)$; $K(-)$, $P(-)$

Let

$K(-)$ be within the square of $P(+)$;

$K(+)$ be no further than $K(-)$ is from $P(+)$ and from its promotion rank;

$P(-)$ be not on any file between $K(+)$ and $P(+)$, inclusive;

$K(+)$ and $P(+)$ be not on Rook files or Knight files.

```
THE TREE OF THE POSITION NUMBER  1
WHITE *KH8,PC6,
BLACK *KA6,PH5,
 WHITE TO PLAY

PC6-C7
    PH5-H4
            PC7-C8Q
            PC7-C8Q
    PH5-H4
    KA6-B7
            PC7-C8Q
                KB7*C8
                    KH8-G7
                        PH5-H4
                            KG7-F6
                                PH4-H3
                                    KF6-F5
                                        PH3-H2
                                            KF5-F4
                                                PH2-H1Q
                                                PH2-H1Q
                                            KF5-F4    -9
                                        PH3-H2    -9
                                    KF6-F5    -9
                                PH4-H3    -9
                            KG7-F6    -9
                        PH5-H4    -9
                    KH8-G7    -9
                KB7*C8    -9
            PC7-C8Q    -9
    KH8-G7
        KB7*C7
            KG7-G6
                PH5-H4
                    KG6-G5
                        PH4-H3
                            KG5-G4
                                PH3-H2
                                    KG4-G3
                                        PH2-H1Q
                                        PH2-H1Q
                                    KG4-G3    -9
                                PH3-H2    -9
                            KG5-G4    -9
                        PH4-H3    -9
                    KG6-G5    -9
                PH5-H4    -9
            KG7-G6    -9
        KB7*C7    -9
    KH8-G7    -9
    KH8-G8
        KB7*C7
```

Figure 44   Printout of a portion of the search tree for the program solving Reti's problem, leading to the introduction of the curtailing of the variation by the rule of the square.

Then

(a) if P(+) is Queened one half-move ahead of P(−), a draw is guaranteed for the (+) side;
(b) if P(+) is Queened one-half move behind P(−), a draw is guaranteed for the (+) side provided that P(−) is not Queened with a simultaneous check and if the appearance of Q(+) is not annihilated by a forked check. A simple proof of the validity of this criterion is very lengthy, and we therefore omit it.

It is worth noting that the concepts of "being in the square", "being no further away", "not being on a file", "is Queened", "earlier", "Queened with simultaneous check", "annihilated by a forked check", etc. were precisely formulated in the program, and took account of the assignment of the move.

As a result of PIONEER's incorporation of these criteria for breaking off a variation, based on extended rules of the square, the search tree in the solution of the problem by Richard Reti (see Fig. 13) took on a certain "human" profile. Until these rules were introduced, PIONEER developed a senseless search even in conditions where a human player would long ago have broken off the variation (see Fig. 44).

We must note that at times the program broke off a variation after a move by Black although the final score assumed a response by White. We may illustrate this by reference to a position occurring in PIONEER's solution of Reti's problem.

In the position shown in Fig. 45, after 3. …Kb6:c6, a chess player would break off the variation because of the reply 4. Ke5-f4. The program, however, did not need to make this move, since the absolute criterion had already operated. In the same way, in reply to 3. …h4-h3, the chess player

Figure 45   Position from PIONEER's search tree for a problem by Reti: White: Kh8, Pc6; Black: Ka6, Ph5. Draw.

would move 4. Ke5-d6, since a draw is guaranteed by the second sufficiency criterion.

3.21. Conclusion

To sum up, we may say that one of the most important tasks in the development of a chess program modelled on the thought processes of a chess master is the construction of a handbook system for recording historical experience, and the elaboration of algorithms for using this experience.

We have reason to believe that the use of the handbook and outreach methods, together with the breaking off of variations by the library method (as is characteristic of chess masters and PIONEER) can be applied in practical control problems in other areas where, as in chess, inexact search and enumeration problems arise.

APPENDIX 4
An Associative Library of Fragments

A. I. Reznitsky and A. D. Yudin

The handbook method described in Appendix 3 cannot be used effectively in the middle game, since at that stage identical positions arise only rarely. A more suitable method consists in selecting a move in the current position by reference to similar positions that have arisen in the past, i.e. in reasoning by analogy.

This method of obtaining information by associating the analysand position with similar positions in the library and using the library data, is called the associative method, and the corresponding library, which stores the indicators of similarity between positions, is called an associative library.

The possibility of developing chess algorithms embodying the use of historical experience by association was mentioned some time ago by Shannon [2]. As an example, he cited the chess master who

> ...knows hundreds, perhaps thousands, of standard positions, customary combinations, and typical maneuvers, which arise frequently in games. There are, for instance, the standard sacrifices of a Knight at f7 or a Bishop at h7, standard mates, e.g. Philidor's mate, maneuvers connected with forks, promotions, etc. In a given position he considers many similarities to cases he knows, and this directs his thinking toward the study of those variations most likely to succeed.

The associative quality of the chess player's thought appears in his attempts to apply the experience of past situations to the present situation by finding some similarities between them.

We have already noted that in Botvinnik's algorithm the mathematical model of a position is constructed in parallel with the development of the tree of variations. The conservation of computer resources depends strongly on the choice of the direction for the search. If we are in a position similar to one encountered earlier and if our accumulated experience allows us to obtain an advantageous configuration of our pieces, or execute a standard combination or maneuver, a further search for variations is unnecessary, since any we found would be pruned. Thus if the possibility of similar action exists in similar situations, we may use it to shrink the tree. This notion was successfully elaborated within the limits of PIONEER and founded on an algorithm operating in middle-game positions and complex endgame positions, in cooperation with the move-search algorithm in the original position.

The method of fragments was developed in order to formalize the notion of similarity among chess positions; its essence is as follows:

Given a past position in which some specific idea (maneuver, combination) was successfully exploited, we extract some portion (fragment) of it, including those pieces whose presence was essential to the implementation of the idea in question. This fragment is stored in the library, and in the future all positions (arising in the search) containing it are considered similar to the position from which it was extracted. It is assumed that in every such position there is an opportunity for the use of the fundamental idea.

Let us describe the basic elements of the method of fragments.

A fragment includes only those pieces whose absence would make the execution of the typical idea impossible. We call these *operational pieces*.

They may be divided into two groups. *Fixed pieces* must be stationed on uniquely determined squares, called *fixation squares*. Trajectories must exist for the remaining pieces (within a predetermined horizon) to other defined squares, which we call *binding squares*; the pieces are *bound pieces*.

The designation of certain operational pieces may be non-uniquely determined, as for instance when the sole function of the piece is to block a trajectory of another operational piece.

The division of the operational pieces into the fixed and bound categories arises from the fact that only one of the players can be interested in the existence of the typical idea. He may move some of his pieces during the preparation and execution of the idea. Therefore the operational pieces belonging to the active player are said to be bound, and their configuration is characterized by a high degree of freedom. The operational pieces belonging to the opposing player are fixed, their positions being subject only to examination.

We distinguish fragments as being *initial* or *nearly initial*, depending on the distance of the bound pieces in the fragment from the corresponding binding squares. An initial fragment is characterized by a configuration of the bound pieces in which each is at a given distance from its binding square. This distance is prescribed in the description of the fragment, individually for each bound piece, and is called the *minimal separation from the binding square*.

Let n_i be the minimal separation of the ith piece from its binding square (as given in the description of the fragment); let l_i be the actual separation of the ith piece from its binding square (the number of half-moves required to reach the square); and let k be the number of bound pieces in the fragment. Then in order that the position should contain an initial fragment, it is necessary that (1) each fixed piece be on its fixation square and (2) for each $i = 1, 2, \ldots, k$, we have $l_i = n_i$.

Next, let H_L be the limiting horizon and let D be a given integer.

If all the fixed pieces are on their fixation squares, and if the following system of inequalities is satisfied:

$$n_i \le l_i \le H_L, \qquad i = 1, 2, \ldots, k, \qquad \sum_{i=1}^{k} (l_i - n_i) \le D,$$

we say that the position contains a nearly initial fragment.

The sum $\Sigma(l_i - n_i)$ represents the distance between the nearly initial fragment and the corresponding initial fragment, i.e. it is the minimum time, in half-moves, required to reach the initial fragment. We call this sum the depth of association of the fragment in the position being analyzed. For an initial fragment, the depth is zero. The quantity D is called the limiting depth of association and is a constraint on the search. It will vary depending on the extent of the current resources.

The role of the associative method in PIONEER is to allow us to determine the priority of variations in the search, at the control system levels in which the pieces contained in the fragment take part in the play, i.e. it allows us to direct the search. The priority for inclusion of a fragment in the search depends on the value of the target in the field where the play occurs in the fragment (the value is recorded in the description of the fragment) and on the depth of association of the fragment in the position being analyzed. Thus, the fragments supplement the system of priorities for the inclusion of fields in the search, which is a subroutine in the move-search program as applied to the original situation.

In positions containing a fragment near the initial situation, the active side may attempt to reach the initial fragment. Then those bound pieces further away from their binding squares than the minimal distance are singled out; trajectories leading to the binding squares are computed and included in the mathematical model. The fields corresponding to the resultant stem trajectories are included in the play (see Appendix 1). The priorities for the moves in the trajectories included in these fields are determined in accordance with PIONEER's priority system.

The preliminary movement of the pieces along the trajectories of these fields or the trajectories for play in the dependent fields may continue until either the initial position is reached or it becomes clear that for one of the following reasons it cannot be reached:

—the configuration of the fixed pieces has been disturbed;

—a bound piece has been captured;

—the trajectories of some bound piece leading to its binding square cannot be unblocked.

If at least one of these conditions is satisfied, all the fields included for the purpose of organizing the outreach to the initial fragment are erased, which is equivalent to erasure for lack of connection to the operational trajectory (see Appendix 1).

If the initial fragment is included in the search, the piece making the first move is identified; the move "according to the fragment" is given highest priority. This move either reaches the corresponding binding square or passes along the trajectory leading to it if the minimal distance for the selected piece is greater than one half-move.

For these reasons, we record the following information for all initial fragments contained in the library:

(1) for the fixed pieces: the identity of the piece and the location of the fixation square;

(2) for the bound pieces: the identity of the piece, the location of the binding square, and the minimal distance from the binding square;

(3) the bound piece making the first move;

(4) the possible amounts of displacement: a vertical or horizontal simultaneous translation of all fixed and bound pieces is called a displacement;

(5) the value of the target.

The development of a library of fragments is justifiable only if recourse to it allows us to save time on the average in our search for strong moves in those positions that contain library fragments. Therefore one requirement placed on the subroutine that searches for fragments is that the time spent in using it must be small in comparison to the mean time spent in analyzing a position.

To save time in the subroutine, we organized a multi-level search. At each level we test for certain specific conditions, and a further search is undertaken only among those fragments that satisfy them. The levels of search may be regarded as filters that pass only the fragments that the given filter deems capable of being found in the position being analyzed.

Fragments that pass a filter are called *candidates*. Their identifying number in the library is recorded in a list of candidates. At each step in the search some fragments fail to pass the filter, and a new list of survivors is made up. The entire library is subjected to the first filter; the last filter passes only those fragments that exist in the position being analyzed. The filters are arranged in increasing order of time consumed in their application, so that the slowest filter confronts the minimum number of candidate fragments. With this organization, the subroutine can select the fragments present in the position in the shortest possible time.

Let us trace the operation of the subroutine in the flowchart shown in Fig. 46.

The information recorded in the library corresponds to the case in which the presence of the fragment on the board is advantageous to White. The recognition subroutine operates on a defined color and finds only those fragments advantageous to that color. Therefore if it is looking for a fragment in favor of Black, as established by test A, it performs a color symmetry, via the procedure B.

Procedure C reflects the analysand position in the vertical axis of the board, in preparation for a flank-symmetrization. The resultant reflection is then used in the same way as the analysand, i.e. we test for the presence of library fragments in either the position itself or in the reflection. If a fragment is contained in the reflection, the a-squares in the trajectories of the bound pieces are subjected to a flank-symmetrization.

Procedure D is the first filter. Here we test whether all the operational pieces are present on the board. We form the first list of candidates from those fragments that survive the filter.

Control then passes to procedure E, the second filter. We test whether the fixed pieces in the analysand position are in their assigned places. The list of candidates is reduced.

Procedure F organizes the search for the bound pieces, and prepares the final list of candidate fragments, namely those that are present in the analysand position. For each fragment in that list it finds: (1) the a_0- and

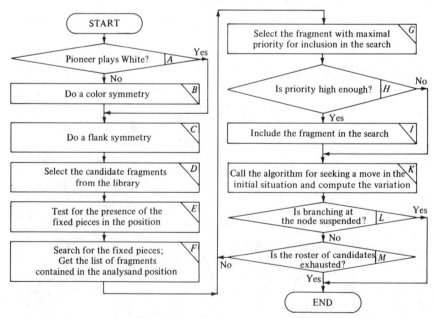

Figure 46   Flowchart of the program for using the associative library of fragments.

a_f-squares in the trajectories of the bound pieces leading from the binding squares to squares more distant than the minimum distance; (2) the depth of association of the fragment in the analysand.

Control passes to procedure G, which selects from the final list of candidates the one with highest priority for inclusion in the search, namely the fragment for which the function $F = (10M - D_a)$ is maximal, where M is the value of the target and D_a is the depth of association of the fragment in the analysand.

After the fragment is chosen, procedure H decides whether to include it in the search, by reference to PIONEER's system of priorities. If play according to the fragment has the highest priority, control passes to procedure I which links the recognition algorithm to the move-search algorithm in the original situation. Here we raise the priority of the "move according to the initial fragment", or if the fragment we have found is nearly initial we include in the play the trajectories passing through the a-squares that we have found. In this way procedure I directs the search.

Control now passes to procedure K, the move-search subroutine in the original situation. We determine the score for the optimal variation. If it is sufficient for breaking off the branching at the node of the search tree where the analysand position arose (procedure L decides this), the library is no longer needed at that node and reference to the library ceases. In the contrary case we see, via procedure M, whether the list of candidate fragments has been exhausted, and if not, control returns to procedure G to test for inclusion of the next fragment on the list. Thus, at some node of the search tree the cycle $G - M$ will be repeated until a decision is reached to terminate the branching there or until the list of fragments in the analysand position is exhausted.

As a typical example, let us consider a combination that has been known for over 300 years. It begins with a Bishop sacrifice on h7. With this sacrifice

Figure 47   Position from the game Schlecter-Wolff.

the attacking side lays bare the position of the King, and then the combined action of Knight and Queen poses mating threats that often cannot be parried.

The position of one side, in which this combination occurs, is shown in Fig. 47. After 16. e4-e5 Nf6-d5; 17. Nc3:d5 e6:d5; 18. Bg5:e7 Nc6:e7, all the conditions have been created for entering the subject combination: 19. Bb1:h7+ Kg8:h7; 20. Nf3-g5+ Kh7-g6. (A retreat by the King 20. ...Kg8 would be followed by 21. Qh5 with an immediate win.) 21. Qd1-g4. White now has a winning position and in a few moves will carry the game through to a win.

Now let us define the initial fragment corresponding to this combination:

1. The fixed pieces:

Name of the Piece	Fixation Square
King	g8
Pawn	g7
Pawn	h7

2. Bound pieces:

Name of the Piece	Binding Square	Minimum Distance from Binding Square
Bishop	h7	1
Knight	g5	1
Queen	h7	2

3. Bound piece making the first move: Bishop.
4. Permissible magnitude of displacement: 0 (displacement not allowed).
5. Value of the target: 200 (value adopted for the King by PIONEER).

Figure 48   A position that might be encountered in an initial fragment.

One of the possible configurations in the initial fragment is shown in Fig. 48. The position shown in Fig. 47 contains a fragment near the initial one. White may reach the initial fragment by unblocking the b-square e4 in the Bishop's trajectory and the a-square g5 in the Knight's trajectory. At least two half-moves are required, and therefore the depth of association of the fragment in the position shown in Fig. 47 is equal to 2. After the moves e4-e5 and Bg5 − e7, White has reached a position containing the initial fragment. The first move according to the fragment, Bb1:h7 + is included in the search and leads to a win.

The method of fragments is able to direct the search for a variation in this way.

In conclusion, we note that the problem of reaching a decision in similar positions is a real one in many control tasks. Methods modelled on the thinking of chess masters will in all likelihood be applicable in various areas of the national economy and will be of a practical value.

References

1. Botvinnik, M. M. Algoritm igry v saxmaty. Moscow: Nauka (1968). (*A Chess Algorithm*. English translation *Chess, Computers, and Long-Range Planning*, Springer-Verlag, Berlin, Heidelberg, and New York (1970))

2. Šennon, K. Raboty po teorii informacii i kibernetike. Sb. Stateĭ. Moscow: IL (1963). (Shannon, C. *Papers on Information Theory and Cybernetics*. Moscow: Foreign Language Press)

3. Botvinnik, M. M. O kibernetičeskoĭ celi igry. Moscow: Sov. radio (1975). (*On the Cybernetic Goal of a Game*)

4. Yudin, A. D. Biblioteka endšpilya EVM. *Šahmaty v SSSR*, 1975, No. 7, 10–11 (An Endgame Library for a Computer)

5. Stilman, B. M. Mašina učitsya. *Šahmaty v SSSR*, 1976, No. 4, 20–22. (The Computer Learns)

6. Yudin, A. D. Programma poiska informacii v dvumernoĭ tablice s subordinacieĭ vhodov. (Biblioteka posiciĭ endšpilya). *Programmirovanie* 1976, No. 4, 66–72. (A Program for Information Search in a Two-Dimensional Table with Subordination of Entries. [A Library of Endgame Positions])

7. Stilman, B. Zwei Arbeiten zum Botvinnik'schen Schachalgorithmus. *Abt. Informatic Universitat Dortmund*, 1976, B.29, 1–55. *See also Dep*. VINITI 3947-76 and 3947a-76/ B. M. Stilman, O programme formirovaniya zony igry. Derevo perebora v zone igry. (A Program for Forming Fields of Play. The Search Tree in a Field of Play.)

8. Adel'son-Velskiĭ, G. M. et al. O programmirovaniya igry Vyčislitel'nyh mašin v šahmaty. *UMN*, 1970, t.25, vyp.2(152), 221–260. (On Programming Chess for Computers.)

9. Kapablanka, H. R. Učebnik šahmatnoĭ igry. Moscow: FIS (1975). (Capablanca, J. R. *A Chess Manual*)

10. Botvinnik, M. M., Stilman, B. M., Yudin, A. D. Iskusstvennyĭ šahmatnyĭ master. *Vestnik AN SSSR*, 1978, No. 4, 82–91. (An Artificial Chess Master).

11. Adel'son-Velskiĭ, G. M., Arlazarov, V. L., Donskoĭ, M. V. Programmirovanie igr. Moscow: Nauka, 1978. (The Programming of Games)

144

Glossary of Terms

Anti-outreach: lowering the priorities of moves in the trajectories included in the MM when their trajectories lead from an advantageous library position (as scored) to a less favorable position.

Array of dimension 15×15 for a given piece: a table of 15×15 squares. In each square a number is inscribed, representing the least number of moves required to reach that square from the central square on an empty board.

Associative thinking, thinking by analogy (in chess): using known types of ideas for setting up situations (positions); this arises in attempts to use historical experience.

Boundary effect: a phenomenon permitting the subdivision of the set of all positions in a given configuration into non-intersecting subsets; the score, and the first move for the position, are invariant over the members of each subset.

Branch pruning
  upon branching in a sheaf of trajectories: cutting off all further consideration of moves in the trajectories of the sheaf, at a given node of the search tree;
  while moving forward: the a priori rejection of certain moves in every position while descending the search tree (see Shannon's Type B method). In PIONEER, these moves are not included in trajectories;
  while backtracking: cutting off branching at a node while ascending the search tree when it is known that further branching there will not change the result of the minimax procedure.

Class: a portion of the library of technical endgames characterized by a defined relationship of material in various positions.

Configuration (a silhouette of the position): characterizes the relative locations of pieces in a position.

145

Control system: a system which gathers information, processes it, and executes the resulting decision.

Decomposition formulae: formulae permitting the development of an arbitrary position in the set corresponding to a position symbol set.

Depth

of association of a nearly initial fragment: the least time, measured in half-moves, in which the fragment can be reached; the distance between the initial fragment and a nearby fragment;

of inclusion of a field: the depth of the node at which the given field is included in a subtree;

of freezing: the depth of the node at which a given trajectory was frozen for lack of connection to an active trajectory;

of a node in a subtree: the distance in half-moves between the initial and given nodes.

Distance of a piece from its binding square

minimal: the length of the shortest trajectory of the bound piece to its binding square in the initial fragment;

real: the same, but in the position being analyzed.

Difference

maximum allowable (D_{max}): the largest number of differences between two positions equal in material that is permitted if the two positions are to be regarded as near each other;

between two positions equal in material (D_{AB}): the sum of the lengths of the trajectories leading from one (initial) position to the other (library) position.

Displacement: the unit for measuring the length of a trajectory.

Erasure of information on fields: erasure of traces of the corresponding sheaves of trajectories, i.e. attaching the corresponding cells to the list of empty cells and closing the list of filled cells.

Field types: attack, blockade, control, retreat, and deblockade.

Field of play: an ensemble of pieces (both Black and White) and their trajectories, that are united in support of, and opposition to, an attacking piece.

Fragment: the ensemble of active pieces in the analysand position that is required for the implementation of a typical idea;

initial: a fragment in which every fixed piece is on its fixation square and every bound piece is at a given distance (in its trajectory) from its binding square;

near the initial fragment: a fragment from which the initial fragment can be reached in a number of moves not exceeding the depth of association;

Freezing a trajectory

for lack of time: an insufficient value of the parameter T_x prohibits movement of a piece in a denial trajectory in a field;

for lack of connection to an active trajectory: movement of a piece in a trajectory is prohibited when the trajectory on which its own depends satisfies one of the following conditions—either the piece has moved in the trajectory, or it has not yet set foot on the trajectory during the variations developed in the search, or the trajectory is already similarly frozen;

because of exit from the a_0-square during the backtrack up the tree: prohibits the movement of a piece in its trajectory when the master trajectory satisfies one of the following conditions—either the piece itself has left its a_0-square during the backtrack, or the trajectory itself is dependent on a similarly frozen trajectory.

Game tree: a graph in which each node (vertex) except one (the initial node or root) has an immediate ancestor.

Germ of a field: the trace of the sheaf of stem trajectories at its a_0-square.

Goal of a game
   exact, in chess: the goal of each side is to mate the enemy King;
   inexact in a model: to win material (piece values: Pawn, 1; Knight, 3; Bishop, 3; Rook, 5; Queen, 9; King, 200).

Going along the trail: a procedure for tracing out a subtree below a given node and gathering information for deciding whether to include the field of play in the given node.

Half-move: the time unit in chess; the time expended by one side in making a move; used in measuring the time of movement in a trajectory and the length of a variation.

Historical experience handbook system: a collection of procedures containing libraries of openings, middle game fragments, technical endgames, rules, and algorithms for using these libraries; the store of knowledge about chess used by PIONEER.

Horizon
   variable (H_x): the time in half-moves leading to control (blockade) of a square in the stem trajectory of a field; it depends on the number of a-squares between the attacking piece and the square to be controlled;
   limiting (H_L): the maximum allowable time in half-moves for a piece to move in its trajectory, under the assumption that until the given place has completed its trajectory none of its compatriots may move; in essence, the horizon prescribes the limit to the length of a trajectory.

Improvement of the outcome of a search: a decision on the possible extension of the MM after analysis of a subtree below a given node, when it is expected that a new current optimal variation (COV) can be obtained, with a higher score.

Inclusion
   of a field in the play: the decision to move pieces in the trajectories of a given field in a given subtree of a search;
   of an initial fragment in the search: including the trajectory of a given bound piece in the MM.

Limiting length of a variation (depth of truncation of the tree): a number conditionally limiting the length of a variation in the search tree (not to be confused with the horizon H_L).

Linked list: a list in which the several elements are arbitrarily located in the computer memory and are connected by pointers; the pointer itself is contained in the element and points to the successor element.

Mathematical model of a position: the ensemble of fields of play corresponding to a position at a given point of the game (or search).

Method

    branch-and-bound (*alpha-beta cutoff*): a method for pruning the search tree by cutting off branches. It uses a subtree equivalent to the non-pruned subtree but retaining the same optimal variation;

    fragments: the method of analogies, based on the notion that in order to implement a typical idea a particular distribution of the pieces must be present in a position. This distribution is called a fragment;

    Shannon's Type A: finding a move by forming a full-width tree of the possible variations in a given position and searching it to a fixed depth, evaluating the variations by a scoring function, and choosing the best variation by a minimax procedure;

    Shannon's Type B: as Type A, but the search tree includes only variations that make sense. The limiting depth of search is significantly increased;

    sighting: a method for finding new trajectories in a field during the search, by noting the sightings (see).

Non-coincidence

    maximum admissible (D_{max}): a number specifying the maximum number of non-coinciding pieces in two positions that are equal in material and that are to be regarded as near each other;

    of Pawn templates: a fact allowing the exclusion of a whole set of positions from a given configuration in a list of candidates for outreach.

    of positions equal in material: the number of non-coinciding pieces;

Optimal exchange: the unbroken sequence of captures on a given square, resulting from a minimax procedure applied to all such sequences on the given square.

Outreach

    to a library position: including in the MM a set of planned trajectories for the pieces on the active side such that the corresponding moves lead from the initial position to a library position having a favorable outcome; implemented by the outreach method;

    to a target fragment: including in the MM trajectories leading to the bound squares of those bound pieces whose distances from their binding squares exceed the minimal distance; implemented in the association method.

Parameters

    of a trace: the parameters in a thirteen-element list that make up the trace of a sheaf of trajectories, giving information about the role of the sheaf in the mathematical model;

    of a column: the parameters in an eight-element list giving information about a given node in the search tree.

Piece

    active: a piece and its corresponding configuration that must occur in the analysand position, if a typical idea is to be implemented;

    a_0-*piece*: stands on the initial square of its trajectory;

    a_f-*piece*: stands on the terminal square of a stem trajectory;

    bound: an active piece that must have a trajectory of given length, not exceeding the limiting horizon, leading to its binding square;

denial: a piece that has a denial trajectory;

fixed: an active piece that must be on its fixation square;

stem: a piece that has the stem trajectory in an attack field.

Position

analysand (the initial position): a position in the game or in a variation during the search, presented to the program for analysis;

candidate for outreach: a library position that may at a given stage of the search become the target of a subsequently organized outreach from the initial position;

equal in material to the initial position: having exactly the same set of pieces and Pawns (the template) as the initial position;

library: a position contained in the library of technical endgames, either explicitly or by means of a decomposition formula;

near the initial position: a position equal to the initial one in material and satisfying the conditions that the maximum allowable non-coincidence (s_{max}) and the maximum allowable distance (D_{max}) are not exceeded;

symbol: representative of the set of positions in a given configuration; a position explicitly contained in the library of technical endgames, from which any position in the set can be obtained by a decomposition formula.

Positional value

The ratio K_w/K_b, where K_w and K_b are the numbers of squares (in trajectories) controlled by White and Black, respectively.

Principle

of expectation: every possibility is explored only so long as there is hope of improvement (attaining the goal);

of maximum gain: a new possibility is taken into account only if it offers a gain greater than that offered by the possibilities already considered;

of timeliness: a possibility is considered only if the corresponding objects will have time to take part in the play.

Priority: the order of inclusion of moves in the search tree. It is defined by a linear function of several variables.

Prognosis: the optimal variation, predicting the extent to which the goal of an inexact game can be attained.

Pseudosearch: backtracking without a score, along a branch in the tree of the current variation. It is carried out only for the possible inclusion of new trajectories of the field in the search.

Recording a sighting: recording information about the trajectory corresponding to a sighting, i.e. recording the trace of the sheaf at its a_0-square, and in particular recording the germ of a field.

Rule for breaking off a variation: a criterion, contained in a library of rules, which allows one to break off a variation during a search, with an exact score.

Search for a decision in a search task

in the initial situation: the search for a decision by forming a search tree;

by association (analogy): the search for a decision by the directed formation of a tree when the initial situation is like one that had a favorable outcome in the past;

by outreach: the search for a decision by the directed formation of a tree when there is hope of attaining exactly a situation that had no favorable outcome in the past;

by the handbook method: the search for a decision without forming a search tree, when the initial situation is precisely the same as one found in the past and for which the score is known.

Scoring function: a function which assigns to every position a real number, called the score; the scores of the final positions in the several variations in a search are used to compare the variations by the minimax procedure.

Sheaf of trajectories

from an a_0-square to a_f-square, of given length and for a given piece: the ensemble of trajectories leading from the a_0-square to the a_f-square, with the provision that the number of displacements of the piece on any of the trajectories, on a free board, does not exceed the length of the sheaf;

retreat or deblockade: the set of all one-displacement trajectories of the given piece.

Sighting: the existence of a trajectory leading from the current location of a piece of a given color to a square containing a piece of the opposite color. When a piece is moved during a search new sightings arise.

Square

a-square of a trajectory: one on which a piece moving in the trajectory comes to rest;

a_0-square: the initial square of the trajectory;

a_f-square: the final (terminal) square of the trajectory;

b-square: one on which the moving piece does not halt;

binding square: the terminal square in the trajectory of a bound piece;

fixation square: the square on which a fixed piece must remain.

Subsystem

first level: a piece and its trajectories;

second level: a field of play

third level: the mathematical model; the complete system.

Symmetry

color: a transformation reflecting the position in the horizontal axis of the board, so that the colors are interchanged;

diagonal: a transformation reflecting the position in either long diagonal a1-h8 or h1-a8. It is applied to positions without Pawns;

flank: a transformation reflecting the position in the vertical axis of the board.

Target: the a_f-piece in an attack field;

strikable: a vulnerable target for which the stem trajectory consists of a single move and the move is with the attacking side;

vulnerable: a target in an attack field when all the a-squares in the stem trajectory are under the control of the attacking side and so also are all the retreat trajectories of the a_f-piece.

Task (problem)

inexact: an enumerative task not soluble exactly but soluble approximately by forming a restricted search tree in connection with an inexact goal of the task;

exact: an enumerative task soluble by using a full-width search tree and an exact goal.

Template

Pawn: an array containing (in general) eight elements, characterizing the distribution of Pawns in a position;

position: an array of twelve elements, characterizing the position as to material.

Trace of a sheaf

A cell in a linked list, attached to a given square, and containing information about some sheaf of trajectories passing through the given square.

Trajectory

denial: a non-stem trajectory of a field; its length is bounded by the variable horizon H_x;

first order denial: connected with the stem trajectory of a field;

n-th order denial: a denial trajectory connected with a denial trajectory of order n-1;

forked: having a portion in common with other trajectories of the same piece;

jointed: consisting of two or more simple trajectories;

planned: moves in this trajectory lead from an initial position to a nearby library position;

shortest (for a given piece, from initial to terminal square): having the fewest moves on a free board among all trajectories leading from the same initial square to the same terminal square;

simple: the shortest, for all pieces except the Queen, for which it is any two-move trajectory on a free board;

stem: the basic trajectory of a field, on which the formation of the field begins; it is limited in length by the limiting horizon H_L.

Translation: the transformation of a position by changing the coordinates of all the fixed and bound squares of the pieces by the same quantity, called the magnitude of the translation.

Two-dimensional table with subordination of entry: a double entry table, one independent and the other dependent on the first.

T_x: the thirteenth parameter in a trace. It directs the search for denial trajectories and one type of freezing; it is measured by the number of moves in a trajectory and depends on the time allotted to a given piece, at a given instant of the play in a field, for movement in a denial trajectory of the field; essentially, it distributes the time allowed by the variable horizon H_x among the denial trajectories of various orders.

Typical idea: a maneuver or combination that has been successful in the past.

Unfreezing a trajectory: the inverse of freezing; to undo the freezing of a trajectory of the same type, on return to the node where it was frozen.

Unravelling a sheaf: obtaining complete information about a sheaf from its trace.

Index of Notation

Index

155